ADVANCES IN UNCONVENTIONAL LITHOGRAPHY

Edited by **Gorgi Kostovski**

Advances in Unconventional Lithography
Edited by Gorgi Kostovski

Published by InTech
Janeza Trdine 9, 51000 Rijeka, Croatia

Publishing Process Manager Alenka Urbancic

Technical Editor Teodora Smiljanic

Cover Designer InTech Design Team

Additional hard copies can be obtained from orders@intechopen.com

Advances in Unconventional Lithography, Edited by Gorgi Kostovski
p. cm.
ISBN 978-953-307-607-2

For EU product safety concerns:
IN TECH d.o.o., Prolaz Marije Krucifikse Kozulić 3, 51000 Rijeka, Croatia,
info@intechopen.com or visit our website at intechopen.com.

Meet the editor

Dr. Gorgi Kostovski received his Ph.D. degree in 2008, following undergraduate degrees in applied physics and communication engineering in 2003, all from RMIT University in Melbourne, Australia.
He is currently a Research Fellow at RMIT, working in the field of nanofabrication for photonics. His research interests include novel nanoscale lithographic strategies for application to low cost fabrication, nanophotonics and optical fibres. His work has been featured on the cover of the prestigious journal Advanced Materials, as well as in Australian media outlets The Age and COSMOS, a popular science magazine.

Contents

Preface

The term Lithography encompasses a range of contemporary technologies for micro and nano scale fabrication. Originally driven by evolution of the semiconductor industry, lithography has grown from its optical origins to demonstrate increasingly fine resolution and to permeate many diverse research fields. Today, greater flexibility and affordability is demanded from lithography than ever before. Diverse needs across many disciplines have produced a multitude of innovative new lithography techniques. This book provides an overview of some of the varied approaches and applications of lithography today.

Nano-imprint lithography is regarded a one of the most competitive alternative fabrication technologies available today. It offers exceptional resolution, high throughput and simple execution at very low cost. The first section of this book, by Barbillon, provides an introduction to nano-imprinting by describing the fabrication of nano-structures for plasmonic biosensing.

Soft lithographies, derived from nano-imprint technologies, operate at largely ambient environmental conditions. This makes them inherently compatible with delicate biological materials such as proteins. Section 2 describes the application of soft lithography in the context of neurobiology. The work by Kan et al. investigates techniques for directing the growth of neurons, with the intention of creating complex neural networks of specific geometry. Similarly, Takayama et al. explore techniques for controlling the differentiation of stem cells.

Self-assembly techniques are often employed for their capacity to pattern large substrate areas with nanoscale resolution at a very low cost. Colloids are a well established platform in the self-assembly arena, and have been applied in areas as diverse as photonic crystals and nanoscale plasmonics. In Section 3, sub-micron colloidal particles are employed as nonplanar substrates for the controlled deposition of a second tier of self-assembled nano-structures. Nonckreman et al. pursue this objective by demonstrating the controlled deposition of nano-sized colloids on top of sub-micron colloids, while Li et al. investigate the self-assembled nano-structuring of thin films as they are deposited on top of close-packed colloids.

Diffractive optical elements typically require fabrication of periodic sub-wavelength elements. For the visible spectrum, this typically translates to sub-micron and nanoscale dimensions. In Section 4, Wen et al. investigate the design and fabrication of circular Dammann diffractive gratings, while Motogaito et al. demonstrate binary diffractive lenses via electron beam lithography. Finally, Fragouli et al. bring this section to a close by demonstrating very interesting work with photochromic polymers. These are imprinted with grating structures, and then shown to be capable of reversible dimensional changes when illuminated with specific wavelengths. Their application to tuneable wetting and tuneable diffraction is also demonstrated.

Section 5 concludes this book by describing some of the challenges faced by the next generation optical lithography technique. Here, shorter wavelengths, in the extreme ultraviolet regime, are harnessed to enable enhanced nanoscale fabrication.

This book represents the collective efforts and creative talents of many researchers worldwide. On behalf of these authors, and the staff at InTech publishing, I hope that you find these collected works informative, and a pleasure to read.

Gorgi Kostovski
RMIT University
Australia

Part 1

Nanoimprint Lithographies

Soft UV Nanoimprint Lithography: A Tool to Design Plasmonic Nanobiosensors

Grégory Barbillon
Laboratoire Charles Fabry de l'Institut d'Optique - CNRS UMR 8501,
Institut d'Optique Graduate School,
France

1. Introduction

The capability for realizing high density nanostructures over large areas is important for the sensing of chemical and biological molecules based on localized surface plasmon resonance (LSPR) of metallic nanoparticles (Jensen et al., 1999; Barbillon et al., 2008; Faure et al., 2008). To characterize these plasmonic nanosensors on an area of $\sim 100 \times 100\ \mu m^2$ (Barbillon et al., 2009; Anker et al., 2008; Barbillon et al., 2008), extinction spectroscopy measurements are mainly used. In order to study multiple biomolecular interactions on the same surface, very large areas need to be fabricated. Various techniques such as focused ion beam lithography and electron beam lithography are available to design these large surfaces. However, these two techniques are slow to obtain these surfaces. Moreover, charge effect on insulating surface can alter the regularity of the pattern shape. Thus, these techniques will not be suitable for a large scale production. Other lithographic techniques such as extreme UV lithography are also used, but these techniques (fabrication of masks) are expensive and allow with difficulty to realize samples in small quantity. In addition, alternative methods emerged, and among these methods we find soft UV nanoimprint lithography (UV-NIL). The UV-NIL process is fast to realize high density nanostructures, not very expensive and compatible with biological and biochemical applications (Krauss & Chou, 1997). With UV-NIL, samples can be fabricated at room temperature and low pressure. A limiting factor of UV-NIL exists and this factor is the resolution of the fabricated molds (Jung et al., 2006; Austin et al., 2005). Flexible molds of the soft UV-NIL technique were fabricated by cast molding processes, in which an appropriate liquid mold material is deposited on a patterned master mold, followed by optical curing of the material. Moreover, a great homogeneity of patterns is obtained with soft UV-NIL on a large zone. Thus, the purpose of this chapter is to present in details the principle of soft UV-NIL and the results of plasmonic structure fabrication on glass substrates obtained by this technique in order to realize LSPR nanosensors for biological molecules. To finish, a plasmonic sensing of biomolecules is investigated in order to validate the use of soft UV-NIL.

2. Soft UV nanimprint lithography: Principle & steps of fabrication

2.1 Principle of UV-NIL

The principle of soft UV-NIL is illustrated in figure 1 and consists of a UV transparent mold that is used to imprint the desired pattern in the UV sensitive resist. This UV sensitive resist, which is liquid at room temperature, is spin coated on the substrate. The UV transparent stamp is deposited on the substrate with a low pressure between 0 and 1 bar (Hamouda et al., 2009), at room temperature. Cross-linking of the UV sensitive resist is then performed by exposing the sample, for example, to a UV lamp source (Hamamatsu LC8 at 365 nm) ensuring a dose of around 10^5 J/m^2 (Hamouda et al., 2010). Next, the soft stamp is released, leaving the UV sensitive resist patterned. The removal of residual layer of UV sensitive resist is realized by etching using Reactive Ion Etching (RIE) in order to obtain the desired patterns. The technique of soft UV-NIL has several advantages like 3D structure generation, and the fabrication of patterns on non-planar surfaces. Indeed, the main advantages of this technique are the transparent flexible stamp and a low viscosity UV-curable resist. These flexible stamps are typically replicated by molding and curing a polymer from a 3D template. The most used materials for UV transparent flexible stamp fabrication is poly(dimethylsiloxane) PDMS (Barbillon et al., 2010), which exhibits interesting properties like good chemical stability and high optical transparency. The fabrication process of soft UV-NIL is divided (figure 1) in 6 steps: (1) the master mold fabrication, which allows flexible stamp realization, (2) the flexible stamp fabrication, (3) the substrate is coated with a UV-curable resist layer, (4) the soft PDMS stamp deposition on the substrate with a low pressure, (5) curing of the photoresist with UV illumination through the transparent stamp, (6) demolding of the soft stamp. Moreover, the use of low viscosity UV-curable resists allows 3D patterning at low pressure without any heating cycles, and thus, the deformation risk of the soft stamp is minimized.

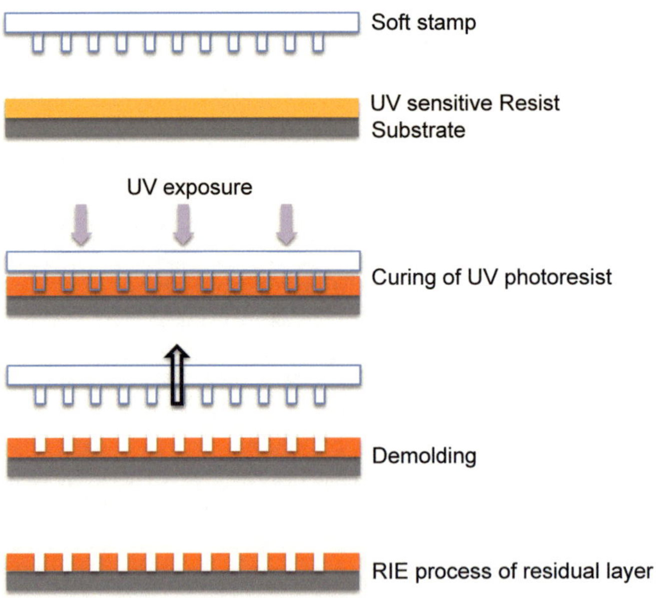

Fig. 1. Principal steps of UV nanoimprint lithography.

2.2 First step of fabrication: Master mold

The method, which is mainly used to fabricate high resolution of nanostructures, is the electron beam lithography (EBL). The advantages of EBL are great accuracy, a very high resolution, and an ability to pattern a large variety of geometries. In the example presented here for Si master mold fabrication, an EBL system (Raith 150) is used to expose the PolyMethylMethAcrylate A6 resist (PMMA A6), employing an accelerating voltage of 20 kV, aperture 7.5 μm and working distance of 7 mm. Next, the patterns designed in PMMA are transferred into the silicon master via a suitable RIE process. The conditions of RIE process are: 10 sccm for O_2, 45 sccm for SF_6 with P = 30 W, a pressure of 50 mTorr and an autopolarization voltage of 85 V (Hamouda et al., 2009; Barbillon et al., 2010). Then, the PMMA mask is removed with acetone. After these steps, the Si master mold surface is treated with HF and H_2O_2 to get a SiO_2 thin surface layer, then modified with an anti-sticking layer (TMCS: TriMethylChloroSilane) to lower the surface energy (Si+TMCS = 28.9 mN/m) which eases the removal of the PDMS molds. In figure 2, SEM images of obtained nanostructures are shown. The chosen geometry for biochemical sensing is nanodisk and consequently, the pattern geometry for the Si master mold designed by EBL is nanohole with following dimensions: diameter of 80 nm and a periodicity of 250 nm.

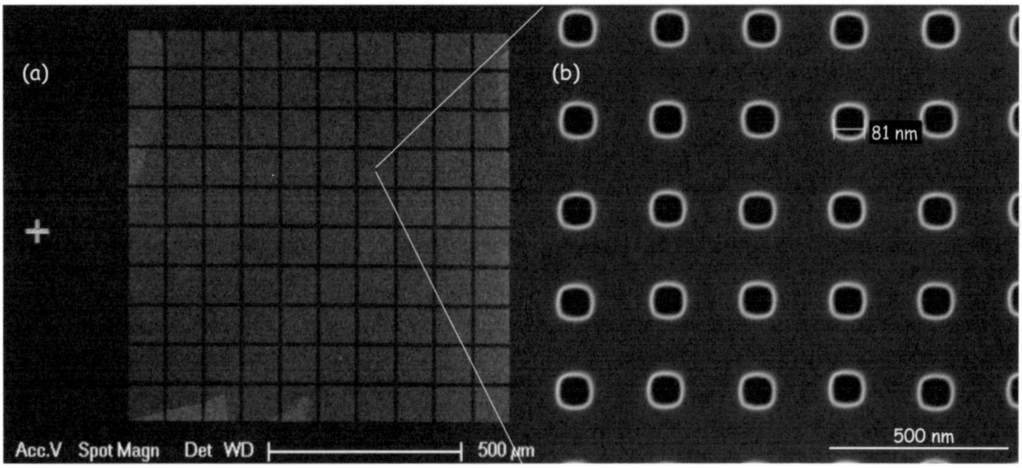

Fig. 2. SEM images of Si master mold designed by EBL: nanoholes of diameter 80 nm and 250 nm of periodicity on a zone of 1 mm^2. (a) the zone of 1 mm^2 and (b) zoom of one square zone where are the nanoholes.

In addition, we developed an alternative method with EBL in order to realize the master mold. This technique is that of nanoporous anodic aluminum oxide (AAO). Nanoporous alumina substrates exhibit an arrangement of nanometric pores, organized in a hexagonal lattice on very large surfaces (Ex: some cm^2). The dimensions of vertical pores can be easily tuned as the diameter, the aspect ratio. The magnitude order of diameter is of 10 to 200 nm, and the aspect ratio can be higher than 500. The synthesis of AAO is realized electrochemically from aluminum wafers. After a step of polishing, anodic potential is applied to an aluminum wafer, at a given and controlled temperature, immersed in an acid bath. The key parameter is the anodization voltage for the growth of these highly ordered nanoporous membranes. In the example that we present here, experimental conditions are chosen in order to obtain membranes with thicknesses of around 10 μm and holes diameter of 180 nm (see figure 3) (Sengupta et al., 2009; Masuda & Fukuda, 1995). Thus, metallic nanodisks will be obtained for plasmonic biosensors application (not shown for this case in this chapter) by using this technique of AAO templates for the UV-sensitive flexible stamp fabrication (Hamouda et al., 2011).

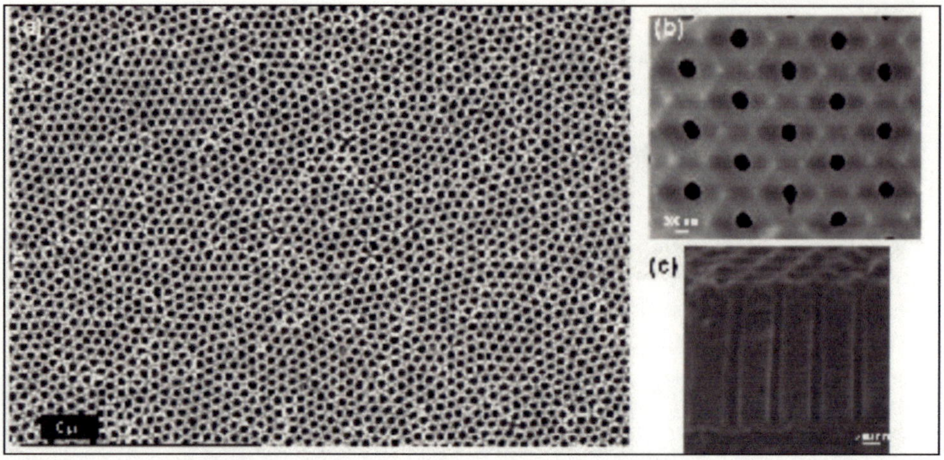

Fig. 3. SEM images of AAO templates: (a) Large view of nanopores, (b) Zoom of these nanopores and (c) Height profile of these vertical nanopores.

2.3 Second step of fabrication: Flexible UV-transparent stamp

For the fabrication of the soft UV-transparent stamp, an elastomer is used on which are realized the desired patterns. The material most used in UV-NIL is poly(dimethylsiloxane) and this material has attractive properties like its low Young's modulus (Bender et al., 2004), its low surface energy, which allows conformal contact with surface without applied pressure and nondestructive release from designed structures (Hsia et al., 2005) and its good transparency to a UV light source (Schmid et al., 1998). Mainly, the used material was undiluted PDMS (RTV 615) and mixed with its curing agent. This mixture was deposited by using a prototype tool and degassed and cured at 75 °C for 12 h (see figure 4). This standard PDMS has some advantages, however a number of properties inherent to PDMS limits its capabilities in the soft UV-NIL. First, the Young's modulus of standard PDMS is low and can limit the fabrication of high density patterns at a sub-100 nm scale due to collapse of structures. Second, the surface energy ($\sim$ 20 mN/m) of PDMS is not low enough to duplicate profiles with high fidelity. To finish, the high elasticity and thermal expansion can lead to deformations and distortions during the fabrication process.

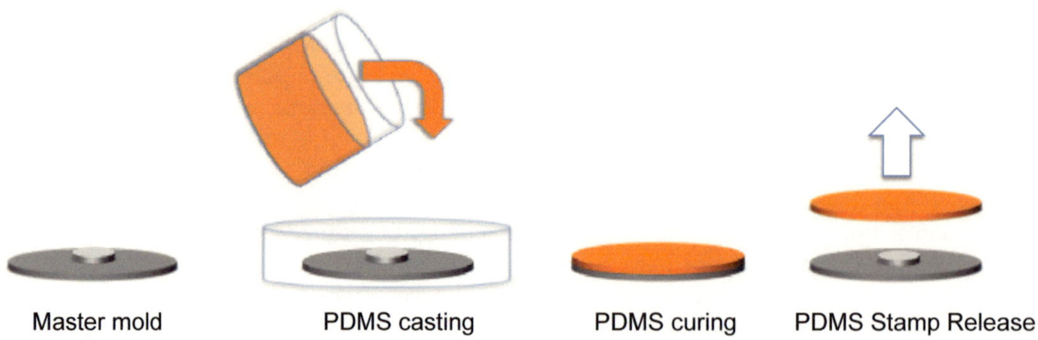

Fig. 4. Scheme of the standard process for the fabrication of the soft stamp.

To improve the resolution and fidelity of structures in the soft UV-NIL, the mechanical properties of the soft stamp need to be improved. Thus, a thin layer of hard PDMS

(50 μm) is used and supported by a standard PDMS layer (1.5 mm) (see figure 5). This second layer allows to keep a good flexibility and adaptation on the spin coated wafer during imprint transfer (Plachetka et al., 2005). Then, the bilayer stamp is fixed on a glass carrier. The hard PDMS is a specific thermocured siloxane polymer based on copolymers Vinylmethylsiloxane-Dimethylsiloxane (VDT301) and Methyl-hydrosilane-Dimethylsiloxane (HMS-301) from ABCR, respectively, 34 g and 11 g (Choi et al., 2004). In addition, before degassing the mixture with a mixing machine we add 50 μL of platinum catalyst, and 0.5% w/w modulator tetramethyl-tetravinyl cyclotetrasiloxane from FLUKA to the mixture (Schmid & Michel, 2000). The hard PDMS is spin coated on the silicon master mold which has been treated with the TriMethylChloroSilane (TMCS) anti-sticking layer. The standard PDMS (RTV 615) with its curing agent are mixed before coating on the thin hard layer PDMS (H-PDMS). Then the sample is cured at 75 °C overnight.

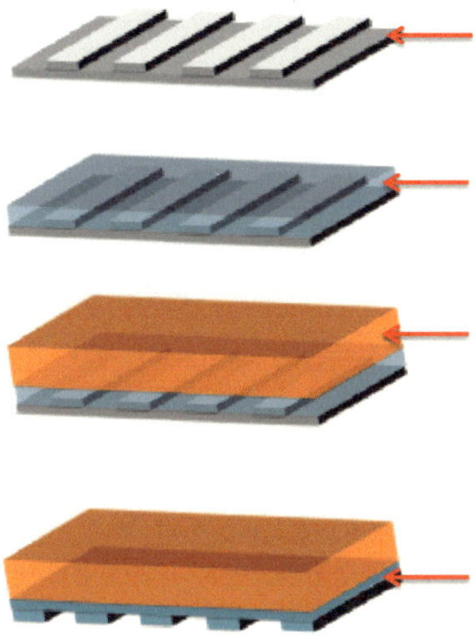

Fig. 5. Principle scheme of the fabrication process of the Hard-PDMS/PDMS stamp.

For the chosen example of nanostructures, this type of soft stamp is very suitable. Indeed, the nanodots dimensions are 80 nm of diameter and the periodicity of 250 nm. The figure 6(a) represents an AFM image of the obtained H-PDMS stamp.

In addition, the SEM image of Fig.6(b) shows nanopatterned hexane diluted PDMS (at 5%, current agent 1/10) surface of around 100 μm^2 by using AAO templates for the fabrication of this soft stamp. The topography of the AAO molds has been well transferred in the PDMS stamp over centimetres of surface area. The stamp is composed of micro-domains of regularly organized nanobumps. Each bump is around 180 nm height and 250 nm diameter. For this fabrication, the PDMS has been simply spread on the AAO membranes without any external pressure but its own weight. Before this spreading, the standard PDMS with the curing agent and the hexane solvent are mixed and cured at 60 °C for 12h. The obtained PDMS layer thickness is around 3 mm. After the curing step, the PDMS layer is peeled off manually. An advantage of using AAO templates is their low surface reactivity compared to silicon dioxide surfaces. Thus, an anti-adhesive layer is not necessary and possible interactions with the used solvent can be avoided, when molding and de-molding PDMS from AAO templates.

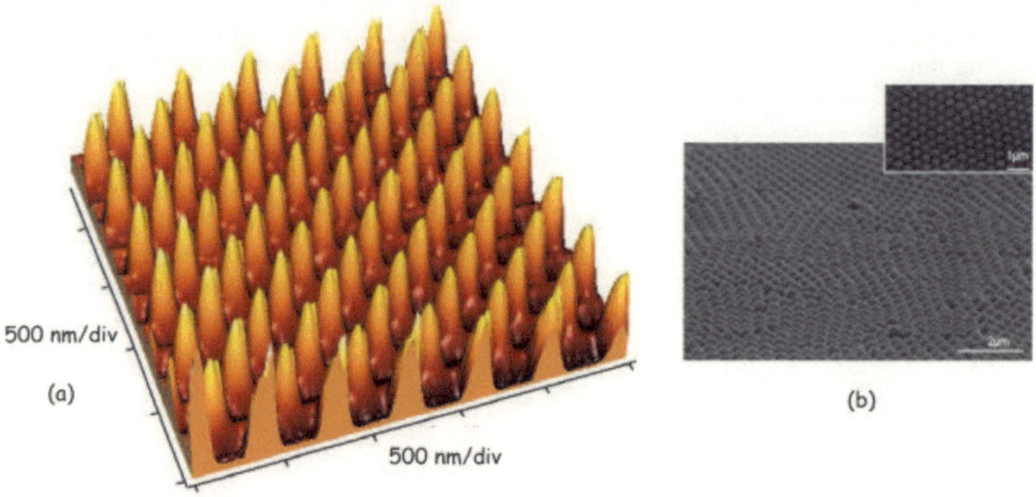

Fig. 6. (a) AFM image of the dots in H-PDMS stamp (periodicity: ~ 250 nm, diameter (FWHM): ~ 80 nm and height: 80 nm), (b) SEM image of stamp of PDMS hexane diluted at 5%, tilted at 55 °, and the insert shows a top view.

2.4 Third step of fabrication: Soft UV-NIL in AMONIL & gold nanodisks fabrication

2.4.1 Soft UV-NIL in AMONIL

Several UV-sensitive resists exist as the NXR 2010 and the AMONIL. Both AMONIL resist and NXR 2010 resist exhibit good performance for resolution and etching resistance. The AMONIL resist was chosen for its low cost compared to the NXR 2010 resist, its excellent time of conservation and AMONIL resist is a mixture of organic and inorganic compounds having a surface energy of 39.5 mN/m. AMONIL MMS4 from AMO GmbH is used and spin coated on the top of a PMMA A2 underlayer (100 nm thick, surface energy = 40.2 mN/m), which is the etching mask for final RIE step and which allows the AMONIL lift-off after curing. For our experiments, an AMONIL thickness of 120 nm is chosen. Then, the imprint process is performed in AMONIL with UV exposure at 365 nm wavelength with 10 mW/cm^2 power during exposure time of 20 min. The pressure used to imprint is 200 mbar. All these parameters were optimized for the fabrication of nanostructures, which use the soft stamp obtained from Si master mold. The figure 7(a) represents the imprint in AMONIL. The dimensions obtained for nanoholes imprinted in AMONIL are ~ 80 nm of diameter and ~ 250 nm of periodicity and these values are in good agreement with the dimensions of nanoholes of Si master mold.

Concerning the soft stamp obtained with the AAO templates, the thicknesses of the PMMA A2 underlayer and AMONIL are 130 nm and 150 nm, respectively. The imprint pressure is achieved by the simple deposit of the PDMS stamp on the top AMONIL layer without any additional (or external) pressure. Only its own weight (~ 1 g) ensures the imprint. For a PDMS stamp of one square centimeter, this would correspond to a pressure of around 70 Pa (Hamouda et al., 2010). In the figure 7(b), the topography of the AMONIL layer after molding is presented. This surface topography is very similar to the AAO master mold. During the imprint, the stamp penetrates completely in the AMONIL layer. The thickness of this layer

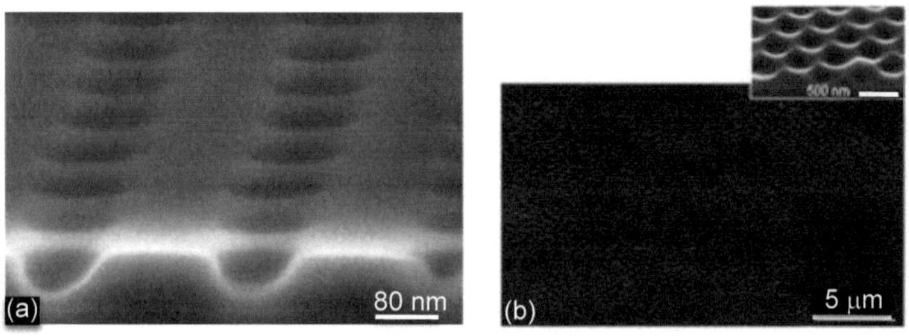

Fig. 7. SEM images of imprint in AMONIL: (a) with soft stamp obtained from Si master mold, and (b) from AAO master mold, the insert is a zoom of imprint of image (b) (scale bar = 500 nm).

is weaker than the as-prepared bump heights. Thus, these results demonstrate the imprint feasibility in AMONIL using the stamps realized with the AAO templates on large zones (cm^2). However, the diameters of nanoholes obtained for the imprint in AMONIL are larger (diameter = $\sim$ 250 nm) than those designed with the AAO master molds. This difference is due to the bad penetration of PDMS in nanoholes. We demonstrated that the penetration is better when the PDMS is well-diluted with hexane (Hamouda et al., 2011). We must still improve the fabrication process of soft stamps.

2.4.2 Gold nanodisks fabrication

Before the fabrication of gold nanodisks, the residual AMONIL thickness in the ground of the nanoholes needs a suitable RIE process before the etch of the PMMA resist. Indeed, this residual thickness is around 30 nm. For the removal of this residual layer, the etch conditions (RIE) are: a flow rate of 2 sccm of O_2, 20 sccm of CHF_3, with a power of P = 25 W, a pressure of 7 mTorr and an autopolarization voltage of 430 V. This gives an etch rate of 18 nm/min in AMONIL. For the removal of the PMMA A2, the conditions of RIE are: 10 sccm of O_2, a power of 10 W, a pressure of 4.7 mTorr and an autopolarization voltage of 280 V. This gives an etch rate of 80 nm/min for PMMA and 18 nm/min for AMONIL. We have a good selectivity between the etch of PMMA and AMONIL. The next step is to evaporate a gold thin layer (50 nm) in order to realize the metallic nanodisks. Previously, an adhesion layer (Cr) for gold is evaporated (3-5 nm). Then, a lift-off in acetone is used to remove the PMMA underlayer (+AMONIL) in order to obtain the gold nanodisks. All these etch, deposition and lift-off conditions are valid for the 2 types of samples. Moreover, an annealing at 250 °C during 30 min for smoothing and compacting the nanodisks. Only the sample of gold nanodisks obtained from Si master mold underwent this annealing. The figure 8 presents the results obtained with the 2 methods of fabrication. We observe that the dimensions of gold nanodisks are in good agreement with the dimensions obtained with the imprint in AMONIL for the 2 methods of fabrication. Indeed, the diameters of gold nanodisks are $\sim$ 80 nm and $\sim$ 250 nm, respectively. On the SEM image 8(a), we note that the annealing has well-smoothed and compacted the gold nanodisks compared to the SEM image 8(b), where the gold nanodisks are not annealed.
Now, gold nanodisks will be used for the plasmonic detection of biomolecules. Only the gold nanodisks of 80 nm diameter will be applied to biodetection. The annealing has slightly changed the shape of gold nanodisks, but no notable change of size was observed. A weak shift is observed (1-3 nm) in the localized surface plasmon resonance (LSPR) wavelength of

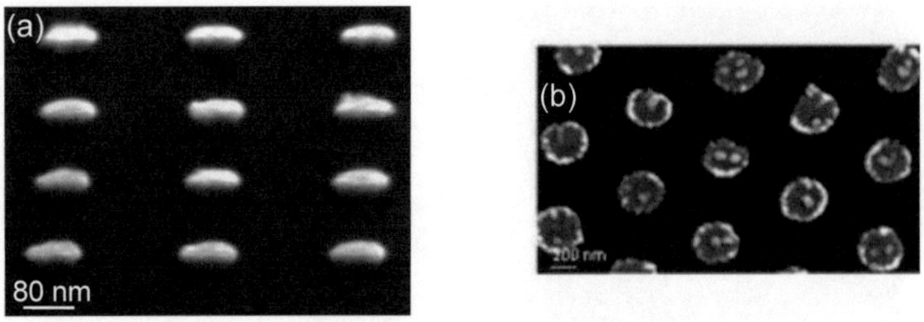

Fig. 8. SEM images of gold nanodisks with the following dimensions: (a) diameter = $\sim$ 80 nm, height = 50 nm and periodicity = $\sim$ 250 nm, and the image tilted at 60 °, (b) diameter = $\sim$ 250 nm, height = 50 nm and the image tilted at 45 °(scale bar = 200 nm).

the gold nanodisks arrays. However, the biomolecule detection will be not affected, because the reference will be taken on the gold nanodisks after annealing.

3. Plasmonic detection of biomolecules

3.1 Adsorbate deposition

To prepare the gold nanodisks for the detection of streptavidin (SA), which is the biomolecule that we chose for its high binding affinity ($K_a \sim 10^{13}\ M^{-1}$) with the biotin molecule, the sample was biotinylated by immersion for 2 h in a solution (1 $mg.mL^{-1}$) of tri-thiolated polypeptides modified with a biotin molecule at their N-term end and washed to remove all unbound molecules. Afterwards, we dried them with N_2 gas. Next, gold nanodisks were incubated in a given concentration of SA for 3 hours. Nanodisks were rinsed thoroughly with 10 mM and 20 mM PBS after biotinylation and after detection of SA to remove non-specifically bound materials. Then, we dried them with N_2 gas.

3.2 Optical characterization of gold plasmonic nanodisks

Visible extinction spectra of gold nanodisks were measured using a Jobin Yvon micro-Raman Spectrometer in standard transmission geometry with unpolarized white light (instrument noise level = $\pm$ 0.5 nm). The sample is located between the white light source and the optical system of detection. The light illuminates the sample under normal incidence and the transmitted light is collected by an objective (x10; N.A = 0.25) on a real area of 30 $\times$ 30 μm^2. The extinction spectra were used to determine the position of the localized surface plasmon resonance of Au nanodisks and the LSPR shift of gold nanodisks after adsorption of molecules (Barbillon et al., 2007; 2008). All measurements were collected in air and to prevent atmospheric contamination, they have been performed on freshly prepared samples. The LSPR wavelength of these gold nanodisk arrays is 615 nm (Figure 9).

3.3 Plasmonic sensing of streptavidin

To validate the fabrication (UV-NIL) of these gold nanodisks and illustrate the properties of these last ones as nanosensors, we have chosen the biotin/streptavidin system. Thus, to evaluate the sensitivity of these gold nanosensors to the detection of this system, a simple model described by Campbell group is used (Jung et al., 1998):

$$\Delta\lambda = m\Delta n \left[1 - \exp\left(\frac{-2d}{l_d}\right)\right] \quad (1)$$

where $\Delta\lambda$ is the wavelength shift, m is the refractive index sensitivity, Δn is the change in refractive index induced by an adsorbate ($\Delta n = n_{adsorbate} - n_{air}$), d is the effective adsorbate layer thickness and l_d is the characteristic evanescent electric field decay length. The sensitivity m does not depend on the adsorbate and according to our measurements (Barbillon et al., 2008; 2009) and those of the Van Duyne group (Jensen et al., 1999), is equal to 2×10^2 nm per refractive index unit (RIU) for our gold nanodisks arrays. To evaluate l_d, the electric field intensity is first calculated by the Finite Difference Time Domain (FDTD) method for different heights from the disk's top and then fitted according to the Prony's method using a single exponential (Barchiesi et al., 2006; Barbillon, 2010). l_d is found equal to 14 nm for gold nanodisks arrays. The index difference between air and Streptavidin is $\Delta n = 0.56$. The size of streptavidin is around 6 nm (Faure et al., 2008).
After the biotinylation, the nanodisks arrays are characterized by an LSPR wavelength at λ_{LSPR} = 621 nm (Figure 9). Figure 9 shows also that a slight redshift of 6 nm is observed after depositing of this biotin layer on gold nanodisks. The sample was then immersed in a PBS (Phosphate Buffered Saline) solution containing the streptavidin at a concentration equal to 1 nM for a duration of 3h and then carefully rinsed. In the figure 9, the extinction spectrum of SA adsorption is represented and the LSPR wavelength was measured at a value of λ_{LSPR} = 644 nm. The real shift due to the presence of SA is then 23 nm.

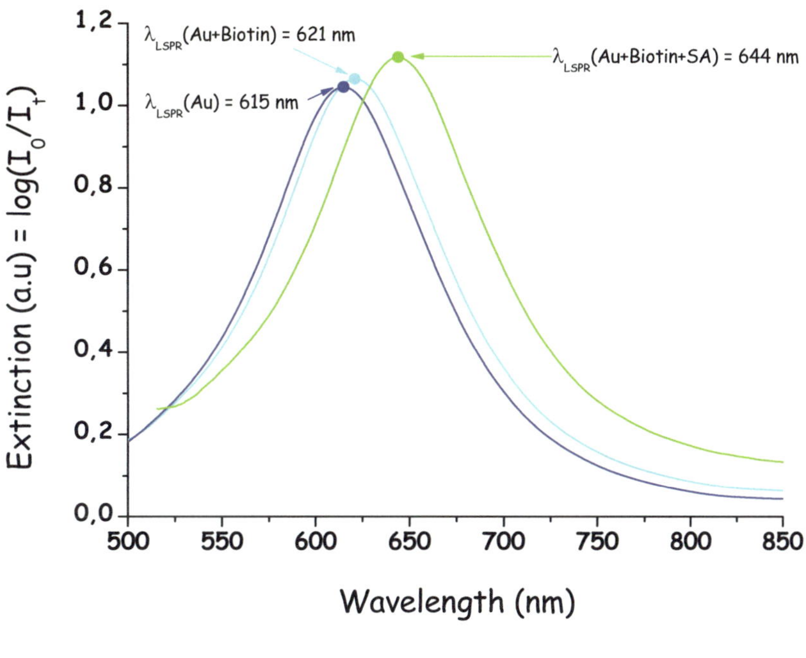

Fig. 9. Extinction spectra at each step of functionalization: in blue, Au before molecule adsorption, in cyan, Au after adsorption of biotin molecules, and in green, Au+Biotin after SA adsorption.

Knowing the following parameters: $\Delta\lambda$, m, Δn and l_d, we could evaluated the value of d and we found d = 1.61 nm. From this value of d, the paving density can be calculated and a value of 0.27 was found. Compared to the maximal density obtained in the case of a hexagonal covering (0.90), the paving density of SA molecules is 30% of the maximal one for the incubation concentration used here (C_{SA} = 1 nM).

4. Conclusion

In this chapter, we demonstrated that the soft UV-NIL technique could fabricated gold nanodisks on large area (Ex: 1 mm^2, and some cm^2). The obtained dimensions of nanodisks are 80 nm of diameter, 250 nm of periodicity and 50 nm of height with the soft stamp designed with Si master mold. In addition, the system AAO and hexane-diluted PDMS demonstrated that it is a good candidate for UV-NIL stamp fabrication, because the advantages of using AAO are to realize inexpensive and efficient hexane-diluted PDMS stamps on large zones (some cm^2). However, the quality of soft stamps realized with AAO must be improved in order to fabricate highly identical and well-defined nanostructures, which could be used for biological applications like bioplasmonics. A small change of the shape of Au nanodisks arrays was also observed after an annealing at 250 °C during 30 min which eliminates the process structural defects. A plasmonic streptavidin sensing with these gold nanodisks was studied. Finally, the gold nanodisks obtained by UV-NIL technique are very sensitive to biomolecules detection. Moreover, we could estimate the paving density of SA adsorbed on gold nanodisks. To finish, the soft UV nanoimprint lithography is very promising and relatively simple to employ for the design of plasmonic nanobiosensors.

5. References

Jensen, T. R.; Duval, M. L.; Kelly, K. L.; Lazarides, A. A.; Schatz, G. C. & Van Duyne, R. P. (1999). Nanosphere Lithography : Effect of the External Dielectric Medium on the Surface Plasmon Resonance Spectrum of a Periodic Array of Silver Nanoparticles. *J. Phys. Chem. B*, 103, 9846-9853, ISSN 1520-5207

Barbillon, G.; Bijeon, J. -L.; Lérondel, G.; Plain, J. & Royer, P. (2008). Detection of chemical molecules with integrated plasmonic glass nanotips. *Surface Science*, 602, L119-L122, ISSN 0039-6028

Faure, A. -C.; Barbillon, G.; Ou, M. G.; Ledoux, G.; Tillement, O.; Roux, S.; Fabregue, D.; Descamps, A.; Bijeon, J. -L.; Marquette, C. A.; Billotey, C.; Jamois, C.; Benyatou, T. & Perriat, P. (2008). Core/Shell nanoparticles for multiple biological detection with enhanced sensitivity and kinetics. *Nanotechnology*, 19, 485103, ISSN 1361-6528

Barbillon, G.; Bijeon, J. -L.; Plain, J. & Royer, P. (2009). Sensitive detection of biological species through localized surface plasmon resonance on gold nanodisks. *Thin Solid Films*, 517, 2997-3000, ISSN 0040-6090

Anker, J. N.; Lyandres, O.; Shah, N. C.; Zhao, J. & Van Duyne, R. P. (2008). Biosensing with plasmonic nanosensors. *Nat. Mater.*, 7, 442-453, ISSN 1476-4660

Barbillon, G.; Faure, A. -C.; El Kork, N.; Moretti, P.; Roux, S.; Tillement, O.; Ou, M. G.; Descamps, A.; Perriat, P.; Vial, A.; Bijeon, J. -L.; Marquette, C. A. & Jacquier, B. (2008). How nanoparticles encapsulating fluorophores allow a double detection of biomolecules by localized surface plasmon resonance and luminescence. *Nanotechnology*, 19, 035705, ISSN 1361-6528

Krauss, P. R. & Chou, S. Y. (1997). Nano-compact disks with 400 Gbit/in^2 storage density fabricated using nanoimprint lithography and real with proximal probe. *Appl. Phys. Lett.*, 71, 3174-3176, ISSN 1077-3118

Jung, G. Y.; Johnston-Halperin, E.; Wu, W.; Yu, Z.; Wang, S. Y.; Tong, W. M.; Li, Z.; Green, J. E.; Sheriff, B. A.; Boukai, A.; Bunimovich, Y.; Heath J. R. & Stanley Williams, R. (2006). Circuit Fabrication at 17 nm Half-Pitch by Nanoimprint Lithography. *Nano Lett.*, 6, 351-354, ISSN 1530-6992

Austin, M. D.; Zhang, W.; Ge, H.; Wasserman, D.; Lyon, S. A. & Chou, S. Y. (2005). 6 nm half-pitch lines and 0.04 μm^2 static random access memory patterns by nanoimprint lithography. *Nanotechnology*, 16, 1058, ISSN 1361-6528

Hamouda, F.; Barbillon, G.; Held, S.; Agnus, G.; Gogol, P.; Maroutian, T.; Scheuring, S. & Bartenlian, B. (2009). Nanoholes by soft UV nanoimprint lithography applied to study of membrane proteins. *Microelectron. Eng.*, 86, 583-585, ISSN 0167-9317

Hamouda, F.; Barbillon, G.; Gaucher, F. & Bartenlian, B. (2010). Sub-200 nm gap electrodes by soft UV nanoimprint lithography using polydimethylsiloxane mold without external pressure. *J. Vac. Sci. Technol. B*, 28, 82-85, ISSN 0003-6951

Barbillon, G.; Hamouda, F.; Held, S.; Gogol, P. & Bartenlian, B. (2010). Gold nanoparticles by soft UV nanoimprint lithography coupled to a lift-off process for plasmonic sensing of antibodies. *Microelectron. Eng.*, 87, 1001-1004, ISSN 0167-9317

Sengupta, K.; Moyen, E.; Macé, M.; Benoliel, A. -M.; Pierres, A.; Thibaudau, F.; Masson, L.; Limozin, L.; Bongrand, P. & Hanbücken, M. (2009). Large-Scale Ordered Plastic Nanopillars for Quantitative Live-Cell Imaging. *Small*, 5, 449-453, ISSN 1613-6829

Masuda, H. & Fukuda, K. (1995). Ordered Metal Nanohole Arrays by a Two-Step Replication of Honeycomb Structures of Anodic Alumina. *Science*, 268, 1466-1468, ISSN 1095-9203

Hamouda, F.; Sahaf, H.; Held, S.; Barbillon, G.; Gogol, P.; Moyen, E.; Aassime, A.; Moreau, J.; Canva, M.; Lourtioz, J. -M.; Hanbücken, M. & Bartenlian, B. (2011). Large area nanopatterning by combined anodic aluminum oxide and soft UV-NIL technologies for applications in biology. *Microelectron. Eng.*, doi: 10.1016/j.mee.2011.02.013, ISSN 0167-9317

Barbillon, G.; Bijeon, J. -L.; Plain, J.; Lamy De La Chapelle, M.; Adam, P. -M. & Royer, P. (2007). Electron beam lithography designed chemical nanosensors based on localized surface plasmon resonance. *Surface Science*, 601, 5057-5061, ISSN 0039-6028

Barbillon, G.; Bijeon, J. -L.; Bouillard, J. -S.; Plain, J.; Lamy De La Chapelle, M.; Adam, P. -M. & Royer, P. (2008). Detection in near-field domain of biomolecules adsorbed on a single metallic nanoparticle. *Journal of Microscopy*, 229, 270-274, ISSN 1365-2818

Bender, M.; Plachteka, U.; Ran, J.; Fuchs, A.; Vratzov, B.; Kurz, H.; Glinsner, T. & Lindner, F. (2004). High resolution lithography with PDMS molds. *J. Vac. Sci. Technol. B*, 22, 3229-3232, ISSN 0003-6951

Hsia, K. J.; Huang, Y.; Menard, E.; Park, J. -U.; Zhou, W.; Rogers, J. & Fulton, J. M. (2005). Collapse of stamps for soft lithography due to interfacial adhesion. *Appl. Phys. Lett.*, 86, 154106, ISSN 0003-6951

Schmid, H.; Biebuyck, H.; Michel, B. & Martin, O. J. M. (1998). Light-coupling masks for lensless, sub-wavelength optical lithography. *Appl. Phys. Lett.*, 72, 2379, ISSN 0003-6951

Plachteka, U.; Bender, M.; Fuchs, A.; Vratzov, B.; Glinsner, T.; Lindner, F. & Kurz, H. (2005). Wafer scale patterning by soft UV-Nanoimprint Lithography. *Microelectron. Eng.*, 73/74, 167-171, ISSN 0167-9317

Choi, D. -G.; Yu, H. -K. & Yang, S. -M. (2004). 2D nano/micro hybrid patterning using soft/block copolymer lithography. *Mater. Sci. Eng. C*, 24, 213-216, ISSN 0928-4931

Schmid, H. & Michel, B. (2000). Siloxane Polymers for High-Resolution, High-Accuracy Soft Lithography. *Macromolecules*, 33, 3042-3049, ISSN 1520-5835

Jung, L. -S.; Campbell, C. -T.; Chinowsky, T.-M.; Mar, M. -N. & Yee, S. -S. (1998). Quantitative Interpretation of the Response of Surface Plasmon Resonance Sensors to Adsorbed Films. *Langmuir*, 14, 5636-5648, ISSN 1520-5827

Barchiesi, D.; Grosges, T. & Vial, A. (2006). Measurement of decay lengths of evanescent waves: the lock-in nonlinear filtering. *New J. Phys.*, 8, 263, ISSN 1367-2630

Barbillon, G. (2010). Determination of Evanescent Electric Field Decay Length of Metallic Nanodisks by Using Localized Surface Plasmon Spectroscopy. *Journal of Materials Science and Engineering*, 4, 69-74, ISSN 1934-8959

Part 2

Lithographies for Biological Applications

2

Application of Soft Lithography and Micro-Fabrication on Neurobiology

Gao Kan, Chen Haifeng, Liu Bing-Fang and Xu Qun-Yuan
Beijing Institute for Neuroscience Beijing Center of Neural Regeneration and Repairing Key Laboratory for Neurodegenerative Diseases of The Ministry Education Capital Medical University Beijing, China

1. Introduction

Soft lithography should be regarded as a complement to common lithography, providing a low-expertise route toward micro/nanofabrication and playing an important role in microfluidics (YN Xia & Whitesides, 1998). The resolution ranges from 5 to 100 nanometer (Pilnam Kim, et al. 2008). Patterns generated by the soft lithography are transfered repeatedly to the soft flexible materials, and then are printed on the medium substrates. In this field, micro contact printing (μCP) is the most widely used technique, especially in bioscience research. Combined with microfluidic patterns technology, several kinds of the extracellular matrix proteins like polymers can be printed to make cells grow according to the designed patterns (Tai Hyun Park, et al. 2003). The cell growth, differentiation in vitro can be regulated in the respect of spatial structure of extracellular matrix(Y. Nam, et al. 2004). So the morphology of neural cells and the influence of spatial structure can be investigated on the micron or even nano-scale level substrates. The closed loop of neural cells can be constructed in order to simulate the complex neural network in vivo. Finally, communication with the specific environment in vitro will be achieved by multi-electrode arrays (MEA). Our previous work used μCP technique can build more solid patterns. By comparing three different extracellular matrixes, PEI can obtain much better results, which adhering more neural cells to form reliable design. However, it is not perfect for the specific neural network construction and the patterned neural cell culture on MEA. In subsequent research, we improved the parameters of the template and achieved a big progress on microfluidic patterning technique to microfabricate patterns. Patterns of biomaterials were constructed with the help of the advanced soft lithography to do the primary cell culture, such as dopaminergic neurons in the substantial nigra and GABAergic neurons in the striatum. Finally, the biocompatibility of MEA was validated initially by primary striatal neuronal culture. Meanwhile, new strategy of structural microfabrication on MEA surface was further explored.

2. Soft lithography materials

Why soft lithography is called "soft"? One of the reasons is the use of soft organic materials such as polydimethyl siloxane(PDMS), polymethyl methacrylate (PMMA), and polycarbonate (PC). PDMS is most widely used in bioscience research, because it has good biocompatibility, chemical stability, optical transparency, air permeability, elasticity. Moreover, the polymer precursors can be aggregated into a mold by UV radiation. PDMS polymerization is shown in Figure 1.

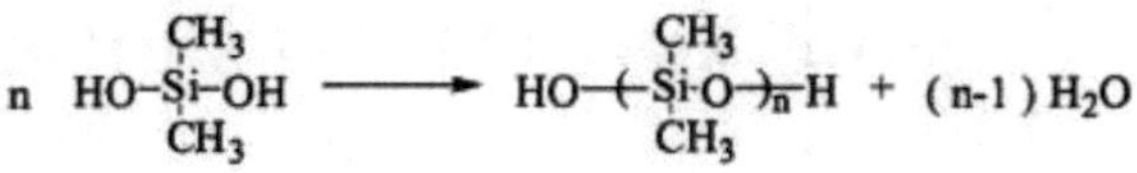

Fig. 1. PDMS polymerization

To cure the PDMS prepolymer in general, a mixture of silicon elastomer and a curing agent (10:1, Sylgard 184 silicone elastomer kit, Dow Corning Corp.) is poured onto the master and placed at 70-80°C for 1 h. The character of the PDMS is closely related to the mixture ratio, curing temperature, and vacuum. Silicon, quartz or glass, and some photoresist are the most common materials to fabricate the masters by standard lithography, transferring the patterns to the PDMS stamp.

3. Soft lithography fabrication methods and it's application in patterning

Various soft lithographic technologies have been applied to fabricate high-quality microstructures and nanostructures including micro contact printing (μCP), replica molding (REM), microtransfer molding (μTM), micromolding in capillaries (MIMIC), and solvent-assisted micromolding (SAMIM). Here, three soft lithographic methods are introduced to fabricate micropatterns onto a surface or MEA: μCP, microfluidic patterning technique and microstencil. The former two can be achieved using the same PDMS stamps and molds. A novel technology was applied to get the high depth-to-width ratio silicon-based mold to fabricate the topographic PDMS microstencil with microfluidic channel (Y.Nam, et al, 2006). Finally, microchannels on MEA with polyimide (PI) guiding the cell growing were also introduced.

3.1 Microcontact printing (μCP)

μCP is a direct method for pattern transfer, generating a non-structured, chemically modified surface. The process of μCP is shown in figure 2. Photolithography was used for the fabrication of silicon-based masters in preparing PDMS stamps. Multi-layer molds were made of thick photoresist like SU-8 on silicon or glass wafers by standard lithography techniques. It was subsequently placed at least 30min in an oven at 160°C to make the photoresist adhere to the substrate closely. Release agent DC20 or OTS were always spin-coated and drying on the master before pouring PDMS. Liquid PDMS (Sylgard-184 from Dow Corning) was poured onto the mold and clamped by the foil, so that the shape of the mold microstructure was transferred to PDMS membrane. It was subsequently placed at least 2 h in an oven at 80°C. The molded PDMS slab was then peeled off and placed onto a glass slide for handling. After curing, PDMS stamps are soaked in a protein "ink", such as PEI, PLL, or LN. 20 minutes later, the "ink" was blew off by using nitrogen gas. Then the raised regions were brought into conformal contact with a substrate in order to print the ink onto the substrate surface. The material of interest was transferred from the PDMS stamp onto the substrate surface. The microscopy of the PDMS stamps were shown in figure 3.

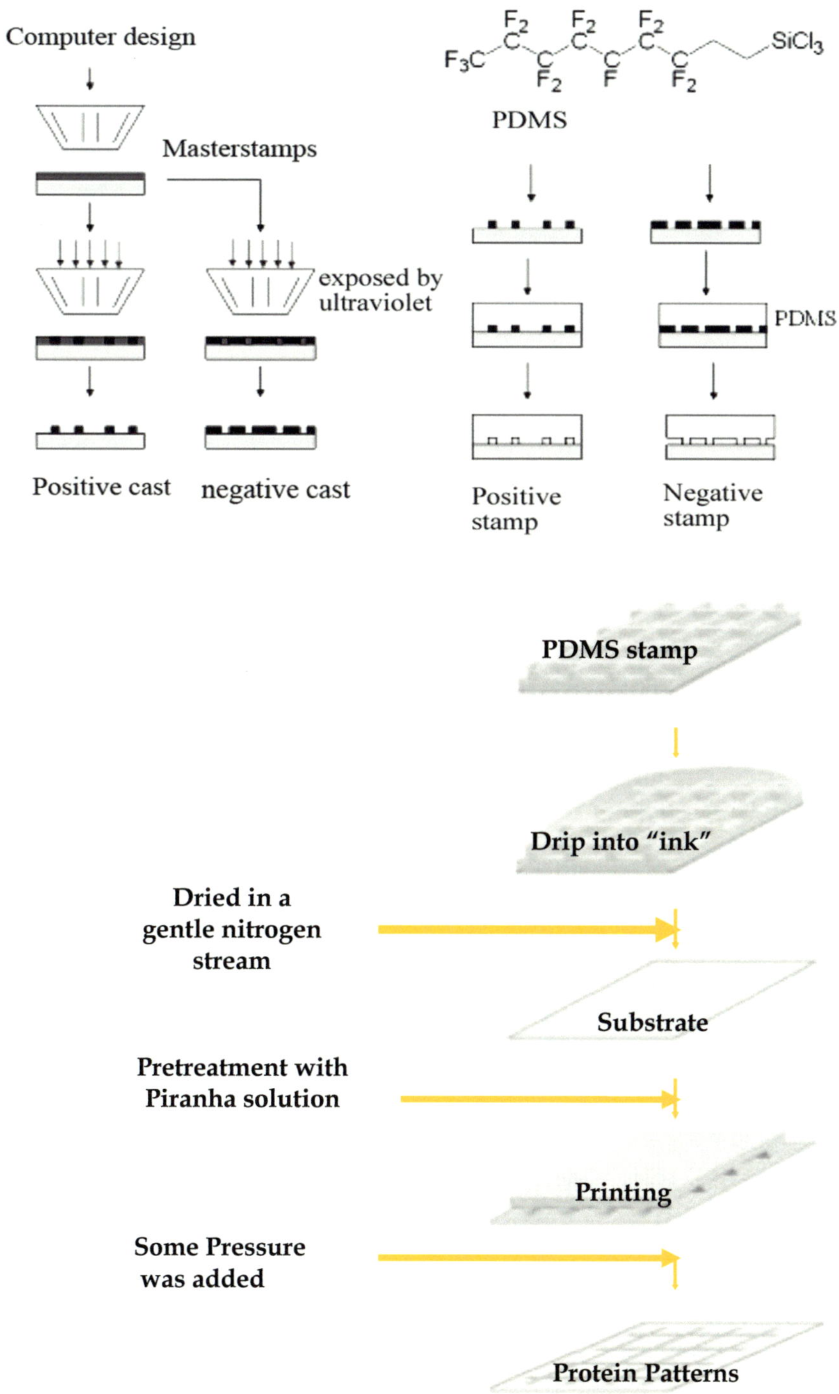

Fig. 2. Schematics of the processes of μCP

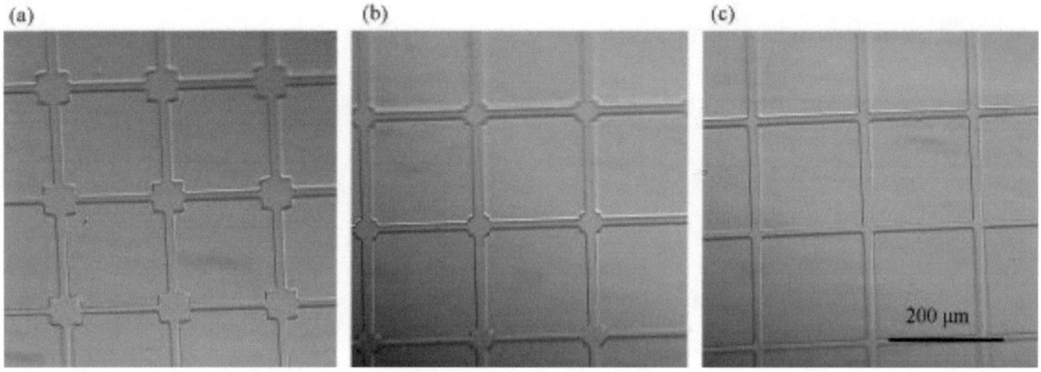

Fig. 3. Microscopy of the PDMS stamps

The bare areas of substrate surface that the PDMS stamp has not touched can be exposed to another coating material. μCP provides the patterning of self-assembled monolayers(SAMs) of alkanethiols on gold, and the resulting control over the adsorption of adhesive proteins facilitates the patterning of cells on substrates.

μCP enables easy stamp replication, fast printing using parallelization, and low-cost batch production. A conformal contact between the stamp and the surface of the substrate is the key to its success. The polymer stamps also minimize the problems of sample carry-over and cross contamination. Printing has the advantage of simplicity and convenience: Once the stamp is available, multiple copies of the pattern can be produced using straightforward experimental techniques. Printing is an additive process; the waste of material is minimized. Printing also has the potential to be used for patterning large areas.

However, μCP has some limitations that are mainly caused by the use of a soft polymer stamp. The swelling of a stamp during inking often results in an increase in the pattern size by diffusion of the excessive printed molecules on the substrate.

3.2 Microfluidic patterning using microchannels

The difference between μCP and microfluidic patterning is that PDMS stamps are soaked in the "ink" in the former usage, but the stamps contact the substrate forming microchannels delivering the materials for cell adhesion or cell suspension to the desired area in the latter usage because of the elastic nature and hydrophobicity of PDMS. The substrate tilted 45 degree, drop of liquids were injected to the PDMS microchannels by the pipette. Then the substrate was put on the test tube rack vertically for 25 to 30 min. Patterns were formed of after the liquid dried.

While this method has been used primarily for surface attachment of cells, it may be possible to adapt this method to three-dimensional tissue constructs. In many cases three dimensional tissue constructs promote cellular differentiation and more authentic cellular morphology and metabolism.

3.3 Microstencil on MEA

The former two methods enable patterning adhesion molecules and guiding cultured cells grow physically. But the cells' communication and interactions in co-cultures are difficult to be detected, which is important to research the function of the cells network.

MEA was a cell-based biosensor for extracellular electrophysiological investigations of neuronal networks. PDMS microstencil was designed to pattern adhesion molecules at the surface of MEA guiding cultured cells grow along the patterns. PDMS microstencil mold was fabricated by a complex photomask aligning method, shown in figure 4.

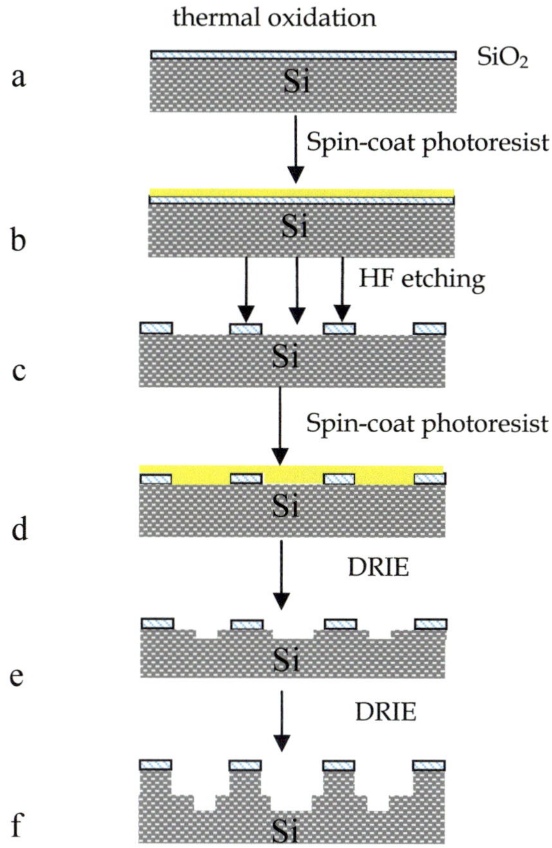

Fig. 4. Schematics of the fabrication of microstencil mold

525μm silicon was thermal oxidated with 4000 Å SiO_2 (see Fig. 4a). The substrate was first coated with a thin photoresist (AZ AZ9912) for 30 s at 3000 rpm (see Fig. 4b). 8 × 8 SiO_2 arrays were fabricated by wet etching as the RIE mask which was the same as the MEA structure (see Fig. 4c). Photoresist was spin-coated to the silicon with SiO_2 mask again (see Fig. 4d) and selectively exposed to UV under a chromium photomask. The silicon was etched 30μm by deep reactive ion etching (DRIE) with photoresist in order to construct the microchannel , (see Fig. 5a). Then photoresist was removed by ultrasonication in acetone. The silicon was selectively etched 70μm by DRIE with SiO_2 resist, forming the topographic PDMS microstencil mold (see Fig. 5b). The high depth-to-width ratio silicon-based mold was designed to penetrate through the PDMS membrane on the MEA to exposure the electrodes and form the microchannel between the electrodes so that the MEA could also work.

(a) (b)

Fig. 5. SEM images of PDMS microstencil mold

The mixture of PDMS prepolymer and curing agent was spin-coated on the mold for 40 s at 4000 rpm. The coated mold was cured for 2 hours at 110∘C in a convection oven. The fully-cured PDMS-coated mold was soaked in an acetone ultra-sonication bath until the PDMS layer released from the mold. The detached microstencil was rinsed with IPA and DI water. The upside and downside of PDMS microstencil with microholes and microchannels SEM images was shown in figure 6.

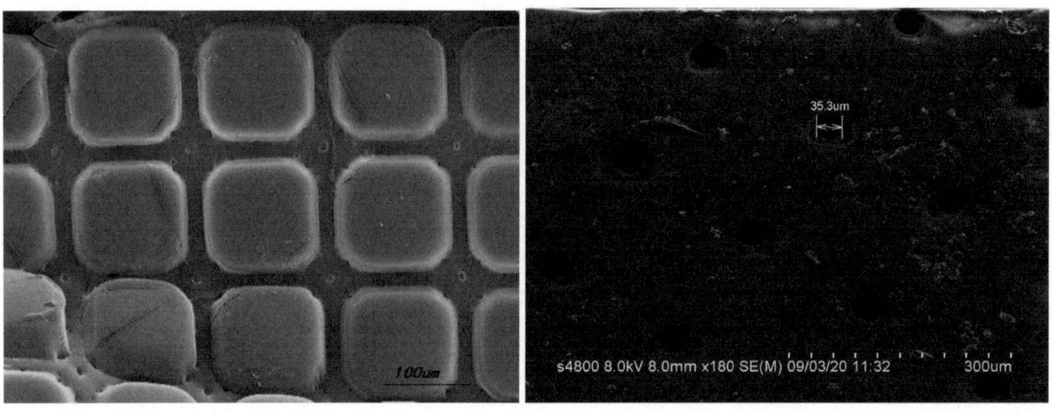

Fig. 6. SEM images of PDMS microstencil

However, PDMS microstencil is difficult to practise because it is hard to align with MEA and the silicon mold is easy to fracture when lifting off.

3.4 MEA with microchannels for patterning

Planar MEA are developed to study electrogenic tissues such as dissociated neuronal cultures (Hiroaki Oka, et al. 1999). They have been widely used with dissociated cultures for a variety of neuroscience investigation including learning and memory and cell-based biosensors for the detection of neurotoxins (Conrad D. James, et al. 2004). But the neurons grow disorderly and cannot form a network so that the function is not the same as the cells in vitro. Combining the patterning technology and MEA forming neuronal networks is the efficient method to research the neurobiology.

μCP, microfluidic patterning technique and microstencil are difficult to operate because the space is too small and the PDMS stamp or stencil can hardly align with MEA. The best and easiest way to forming neuronal networks is to fabricate the microchannels on MEA with polyimide (PI) guiding the cell growing.

MEA were fabricated using a conventional semiconductor process (Guangxin Xiang, et al. 2007). After cleaning the polished quartz glass wafer, the conductive layer of Au/Ti film (Au 3000 Å and Ti 700 Å) was sputtered. 8 × 8 electrode arrays were left with the photomask protection by standard photolithography. Then, a combination of $SiO_2/Si_3N_4/SiO_2$ (3000 Å /4000 Å /3000 Å) passivation layers was deposited onto the substrate using plasma enhanced chemical vapor deposition (PECVD), and the insulating layers on the electrodes and the bonding-pads were removed by inductively coupled plasma (ICP) (see Fig. 7a). Finally, Negative photosensitive polyimide (AP2210B, Fujifilm Electronic Materials Inc) was spin-coated to form microchannels having a thickness of $3\sim4\ \mu m$ and photo-etched by the standard procedure to expose the microelectrodes and the terminals (see Fig. 7b).

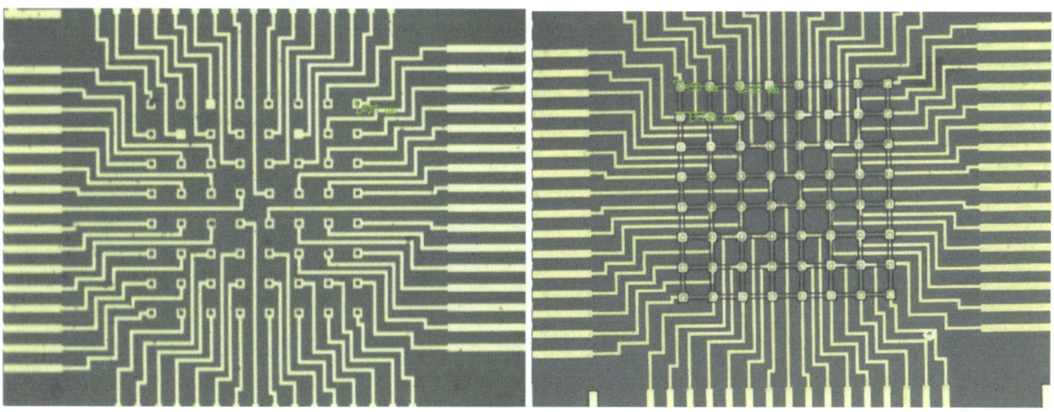

(a) (b)

Fig. 7. Microscopy of the MEA with PI microchannels

GABAergic neurons in the striatum and PC12 cells were cultured on MEA with PI microchannels which were coated with poly-l-lysine (PLL) to promote cell adhesion, (see Fig. 8a, 8b). PI microchannels could be seen between the electrodes and the neural cell can grow along the microchannels. However the nerve cell synapse could not formed along the microchannels. Because the depth of microchannels could not match the neurons and the

PLL could not be guarantee to coat the microchannels effectively after days. There is still a lot of work to do to construct the neuronal networks on MEA to study the cells function as in vitro.

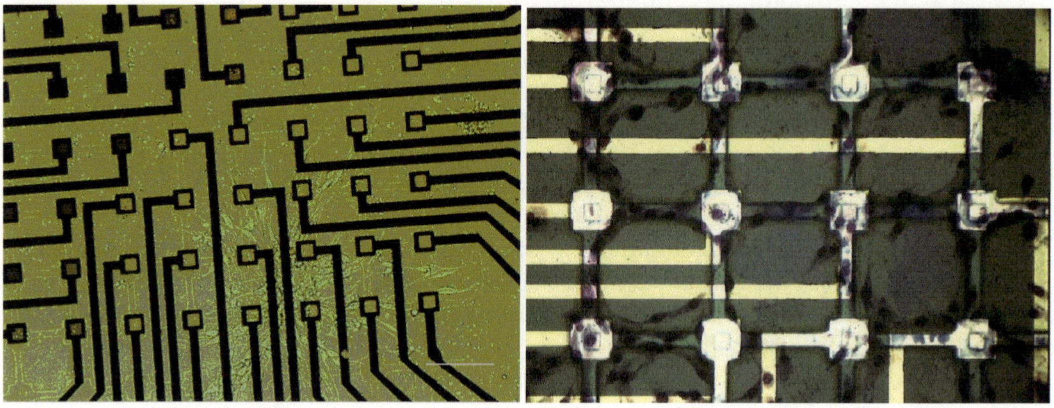

(a) (b)

Fig. 8. Microscopy of the cells cultured on MEA with PI microchannels (a) GABAergic neurons in the striatum cultured at 3 days (b) PC 12 cells cultured at 6 days

4. Construction of neural network by applying soft lithography

A man-made neural network on electrode can be applied to do basic research of neuroscience, be a kind of biosensor for drug discovery [9] and even be implanted into brain to establish artificial connections that could form the basis of a neural prosthesis[10]. These fields have caused much attention in the world [11]. In the past, there are three strategies to realize a simplified neural network in vitro, such as mechanical fixation-applying spatial restrictions[12,13], physical modification of surface roughness and surface topography[14,15], chemical polymer microfabrication on surface—using soft lithography.
Soft lithography is created by Whitesides in 1993[16]. It is used to create chemical structures on surfaces, including μCP, μFN and other downstream techniques. These microfabrication techniques that control both the size and shape of the cell anchored to a particular surface are extremely useful for understanding the influence of the cell–material interface on the behavior of cells [17, 18].The adhesion and survival of neural cells should be considered firstly for the patterned neural cell culture in vitro. So the selection of appropriate cell-attracting substances is an important step for pattern design in micro-contact printing. Ideal substances encourage good cell–substrate interactions, constantly stimulate the cells by substrate-bound chemical, biological, electrical or mechanical signals [19, 20] and even regulate neuritis growth on designed patterns. The most commonly used coating reagents to promote cell adhesion are extracellular matrix (ECM) proteins like laminin (LN), positively charged polymers such as poly-l-lysine (PLL) and synthetic amide-linkage-free compounds such as polyethylenimine (PEI). Therefore, the characteristics of three different substrates, PEI, PLL, LN were compared by the primary neuron culture in our previous work. The PEI characterized with strong positive surface charges was validated to fabricate more continuous and integrated micro-contact printing neural patterns under serum-free culture conditions than PLL, LN[21].

For the functional neural network construction on MEA, an inevitable question that should be addressed finally is how to realize accurate opposite between neurons and the electrode under neurons. We assume firstly microfluidic technique may have more advantages than µCP. In subsequent research, we achieved satisfied patterns by microfluidic technique for further research with the help of the progress on parameters of template. Specific neural network were constructed by applying advanced soft lithography above to do the primary cell culture, such as dopaminergic neurons in the substantial nigra and GABAergic neurons in the striatum. The conditions of neuronal adhesion on different patterns (grids and lines) were also observed using several techniques, including atomic force microscopy, immunohistochemistry, transmission electron microscope and scanning electron microscope.

4.1 Neural network with rat fetal hippocampal cells by µCP patterns

In previous study, we examined the ability of another positively charged polymer, polyethyleneimine (PEI), to promote neuronal adhesion, growth and the formation of a functional neuronal network in vitro. PEI, PLL and LN were used to produce grid-shape patterns on glass coverslips by µCP. Post-mitotic neurons from the rat fetal hippocampus were cultured on the different polymers and the viability and morphology of these neurons under serum-free culture conditions were observed

4.1.1 Cells adhesion

The number of cells that adhere to the different substrates after 24 h in culture is shown in Fig. 9. The adhesive effects were evaluated by calculating the ratio of cell numbers that adhere to the grid-like patterns divided by the total area of printed polymer. We found that the positively charged polymers (PLL and PEI) had a significantly higher level of cell attachment than LN ($p < 0.05$)[13].

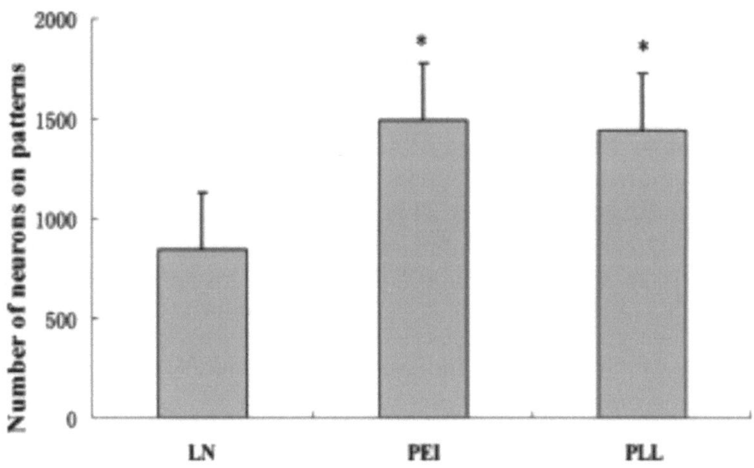

Fig. 9. The adhesive effects were test by analysis of the number of neurons on the area (mm2) of LN, PEI and PLL grid patterns after 24 h in culture. The asterisks indicated neurons on PEI and PLL patterns had significantly higher lever than on LN patterns, n = 12, $p < 0.05$.

4.1.2 Cells viability

In Fig.10, We show that neurons cultured on the PEI- and PLL-coated surfaces adhered to and extended neurites along the grid-shape patterns, whereas neurons cultured on the LN-coated coverslips clustered into clumps of cells. In addition, we found that the neurons on the PEI and PLL-coated grids survived for more than 2 weeks in serum-free conditions, whereas most neurons cultured on the LN-coated grids died after 1 week[13].

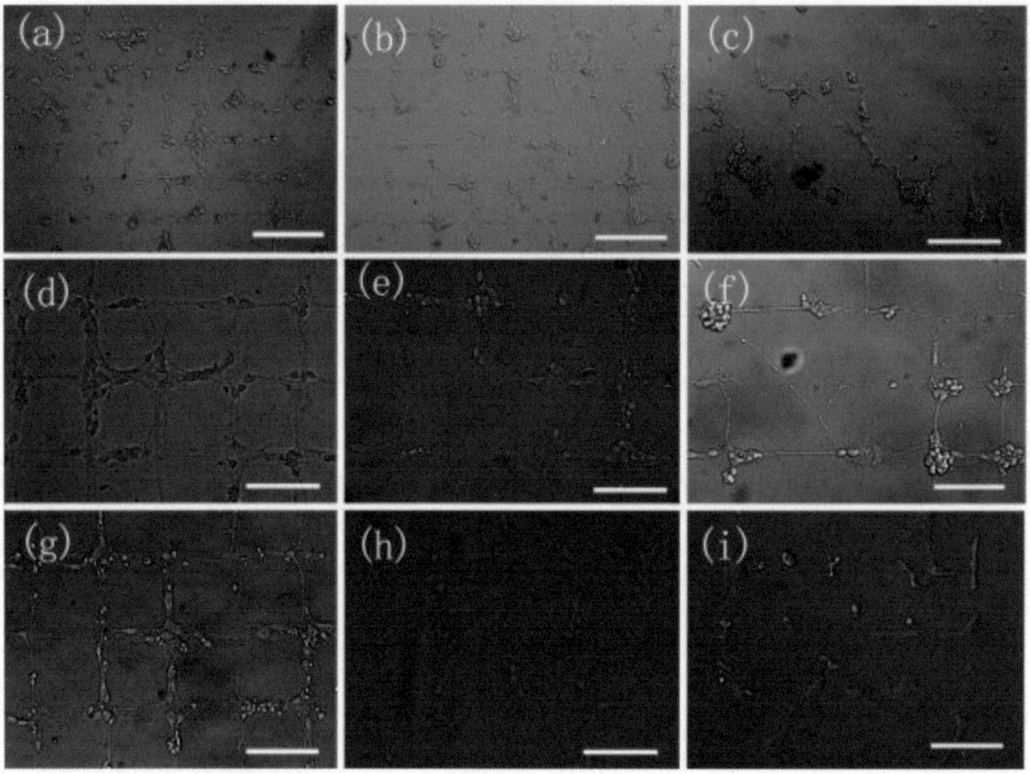

Fig. 10. Images obtained using the phase contrast microscope, showing cells cultured on PEI, PLL and LN polymeric films at different time points. (a, d, and g) Representative images of neurons cultured on PEI grid patterns at 24 h, 7 days, and 14 days, respectively, show that neurons adhere and grow accurately along the PEI grids at differently time points. (b, e, and h) Representative images of neurons cultured on PLL grid patterns at 24 h, 7 days, and 14 days, respectively. (c, f, and i) Representative images of neurons cultured on LN grid patterns at 24 h, 7 days, and 14 days, respectively. The images (c and f) show that the neurons on the LNgrid patterns often accumulate at the cross points of the grids. The image (i) shows that most neurons disappear after 14 days in culture and only small areas of the neural cells exist. (a–c) Bar = 100 μm; (d–i) bar = 50 μm.

4.2 Specific neural network with two relative neurons, such as dopaminergic neurons and GABAergic neurons

There is a closely relationship in the respect of function and structure between dopaminergic neurons in the substantial nigra and GABAergic neurons in the striatum. As well-known,

Parkinson's disease is due to the loss and injuries of dopaminergic neuron in substantia nigra which cause a decrease in nerve fibers projected to the new striatum. Finally, reasonable synaptic connections and neural network can not be established. Therefore, it is expected to establish cell models to investigate the relationship between these two kinds of coherent neurons and construct an artificial neural network in vitro by the application of soft lithography. In present work, specific neural network with dopaminergic neurons and GABAergic neurons co-culture was established by µCP PEI grid patterns. Meanwhile, PEI was validated again to fabricate more continuous and integrated neural patterns by using µCP and microfluidic technique both.

4.2.1 Neural network by µCP

Different kinds of neural network by µCP were established with neuron from the striatum, dopaminergic neurons from the substantial nigra and both of them co-culture. The conditions of neuronal adhesion on different pattern figures were observed using several techniques, including immunocytochemical staining, transmission electron microscope and scanning electron microscope. Using immunocytochemical staining, transmission electron microscope, we identified the types of neural cells and observed some neurosynapse-like structures near the neuronal soma on PEI-coated coverslips. These findings indicate that PEI is a suitable surface for establishing a functional neuronal network in vitro.

4.2.1.1 Investigation of neural cell types and neurite elongation along the grid-like patterns by immunocytochemical staining and SEM

PEI, PLL and LN were used to produce grid-shape patterns on glass coverslips by micro-contact printing. GABAergic neurons and medium spiny neuron from the rat striatum, dopaminergic neurons from the rat substantial nigra and both of them co-culture were researched separately on the different polymers coated surface. The viability and morphology of these neurons under serum-free culture conditions were observed using fluorescent microscopy in Fig. 11, Fig. 13, Fig. 14. After 7 days in culture, we found that the neural cell bodies on the PEI patterns were located mostly at the cross-points of the grid, whereas neurites extended along the line of the grid-like patterns. More continuous and integrated neural network was achieved finally. On the PLL-coated coverslips, the neural patterns appeared to be integrated. But several cells clustered at the cross-points of the grid disappeared gradually after the media was replaced. In contrast, cells cultured on the LN-coated grids were generally clustered into clumps and cannot form satisfied patterns. Different sizes of PEI pattern were produced by microcontact printing. In Fig. 12, compared with 50 μm, 100 μm, 200 μm pattern sizes, we found few difference early. After 7 days or 14 days culture, most neural cells on 200 μm size grew well and were seldom found to overlap each other, unlike those on 50 μm size clustered into clumps at the cross-points of grid and disappeared gradually. Identified with immunocytochemical staining, we found that neural cells from the rat substantial nigra were TH positive, synaptic vesicle protein were synaptophysin positive in Fig. 13 and cells from the rat striatum were GABA positive or DARPP-32 positive in Fig. 11, Fig. 12, Fig. 14. SEM show neurons outgrowth on PEI-coated patterns and validate the findings by immunocytochemical staining in Fig. 15, Fig. 16.

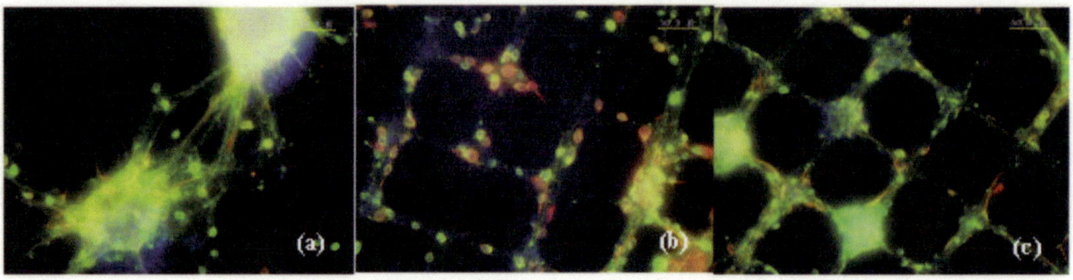

(a) on LN grid patterns (b) on PEI 100 μm grid patterns (c) on PLL 100 μm grid patterns bar=50 μm

Fig. 11. Immunofluorescent image of anti-GABA (green fluorescence) +anti-MAP2 (red fluorescence) labelled striatal neurons cultured for 7 days on different substrates, the nuclei of neurons were stained with Hochest x400

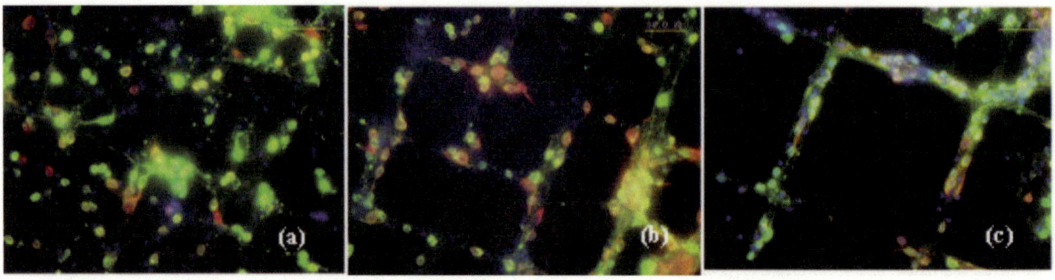

(a)on 50 μm patterns (b) on 100 μm patterns (c) on 200 μm patterns bar=50 μm

Fig. 12. Immunofluorescent image of anti-GABA(green fluorescence)+anti-MAP2(red fluorescence) labeled striatal neurons cultured for 7 days on different sizes of PEI patterns, the nuclei of neurons were stained with Hochest x400

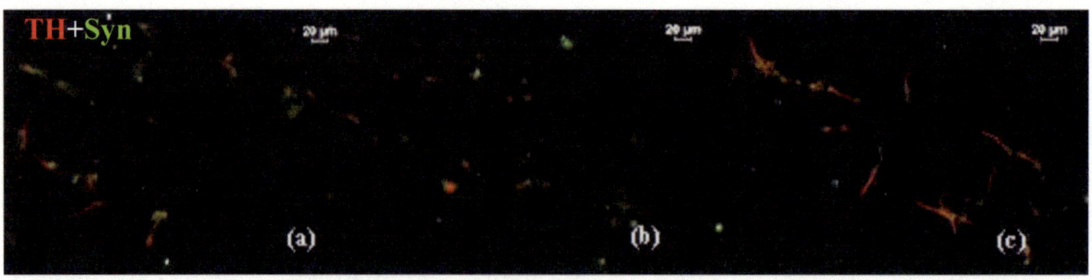

(a) on LN grid patterns (b) on PEI 100 μm grid patterns (c) on LN+PEI bar=20 μm

Fig. 13. Immunofluorescent image of anti-TH(red fluorescence) labeled dopaminergic neurons from the substantial nigra and anti-synaptophysin(green fluorescence) labeled synaptic vesicle protein cultured for 7 days on different substrates. x200

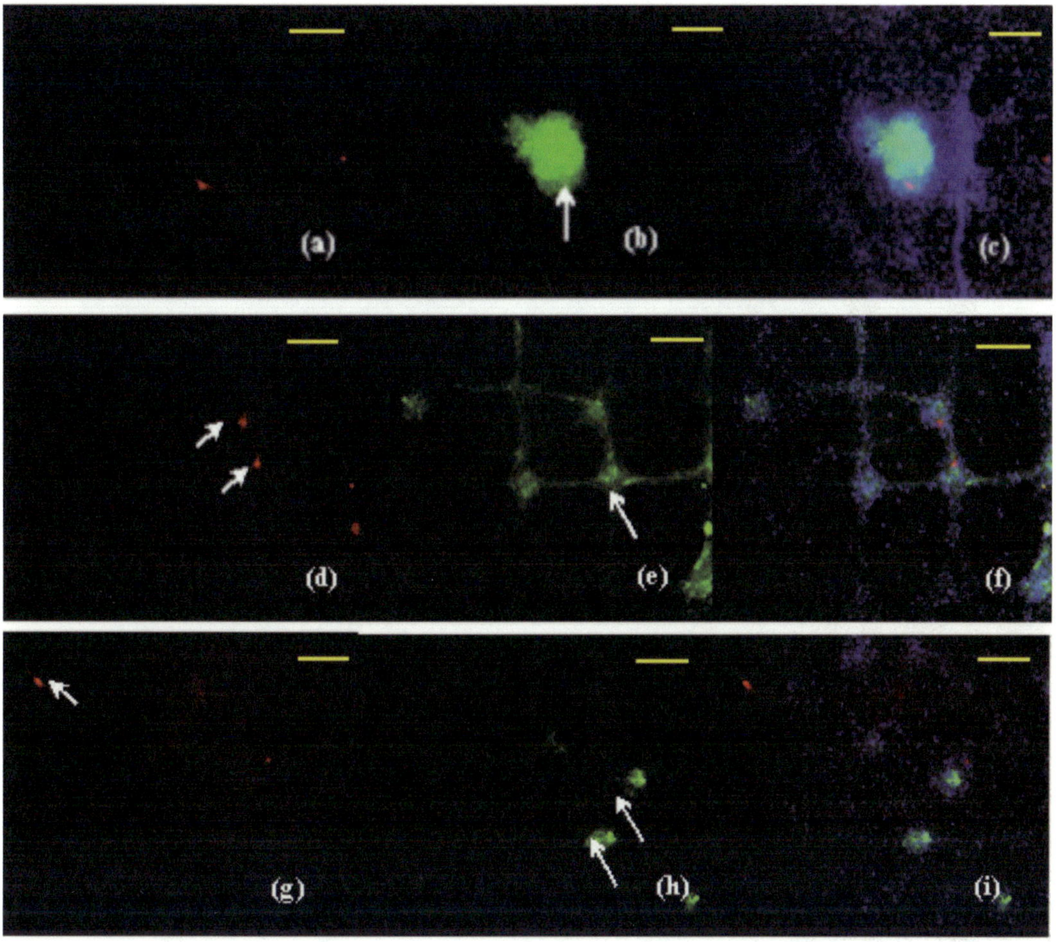

(a–c) co-culture neurons growing on LN grid patterns. (a) Immunostaining with anti-TH (red fluorescence) labelled dopaminergic neurons. (b) Immunostaining with anti-DARPP-32 (green fluorescence) labeled medium spiny neuron from the striatum(arrow). (c) Merged image of (a) and (b), showing TH positive neurons on the LN-coated patterns actually adhere to the cluster formed by medium spiny neuron. (d–f) co-culture neurons growing on PEI grid patterns. (d) Immunostaining with anti-TH (red fluorescence) labelled dopaminergic neurons(arrows). (e) Immunostaining with anti-DARPP-32 (green fluorescence) labeled medium spiny neuron from the striatum(arrow). (f) Merged image of (d) and (e) showing that DARPP32 positive neurons cultured on the PEI-coated patterns form a continuous and integrated neural network, and two TH positive neurons adhere to the cross-points of grid. (g–i) co-culture neurons growing on PLL grid patterns. (g) Immunostaining with anti-TH (red fluorescence) labeled dopaminergic neurons(arrow). (h) Immunostaining with anti-DARPP-32 (green fluorescence) labeled medium spiny neuron from the striatum(arrows). (i) Merged image of (g) and (h)showing only two DARPP32 positive neurons adhere to the cross-points of grid. TH positive neurons were not restricted by the grid pattern. bar=100 μ m

Fig. 14. Immunofluorescent image of anti-TH labelled dopaminergic neurons from the substantial nigra and anti-DARPP32 labeled medium spiny neuron from the striatum co-culture for 7days on different substrates, the nuclei of neurons were stained with Hochest x200

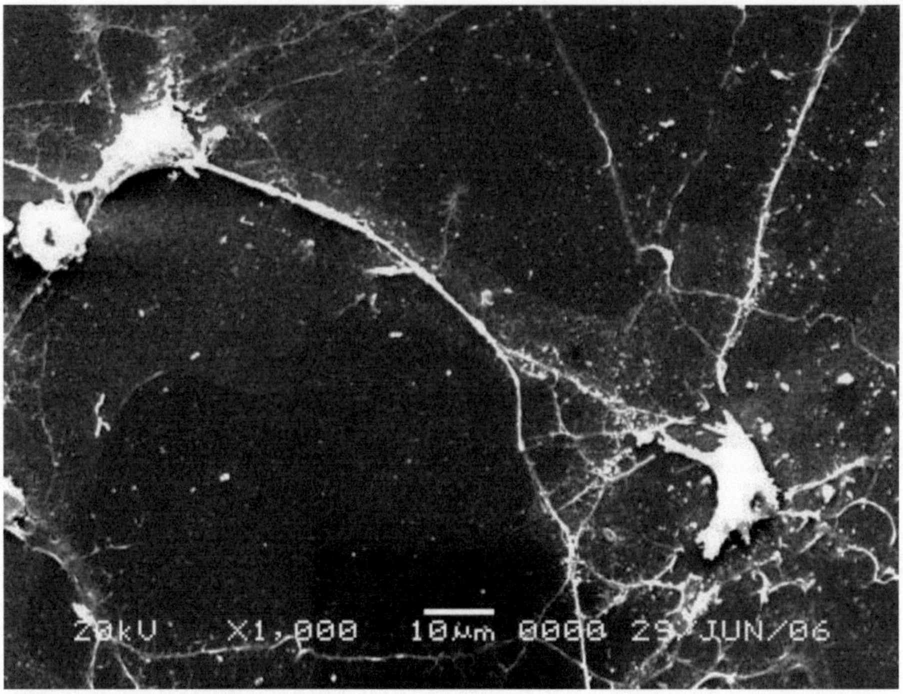

Fig. 15. SEM show neurons outgrowth on PEI-coated patterns. Somata of neuons located on the cross point and neurites extend along the lines

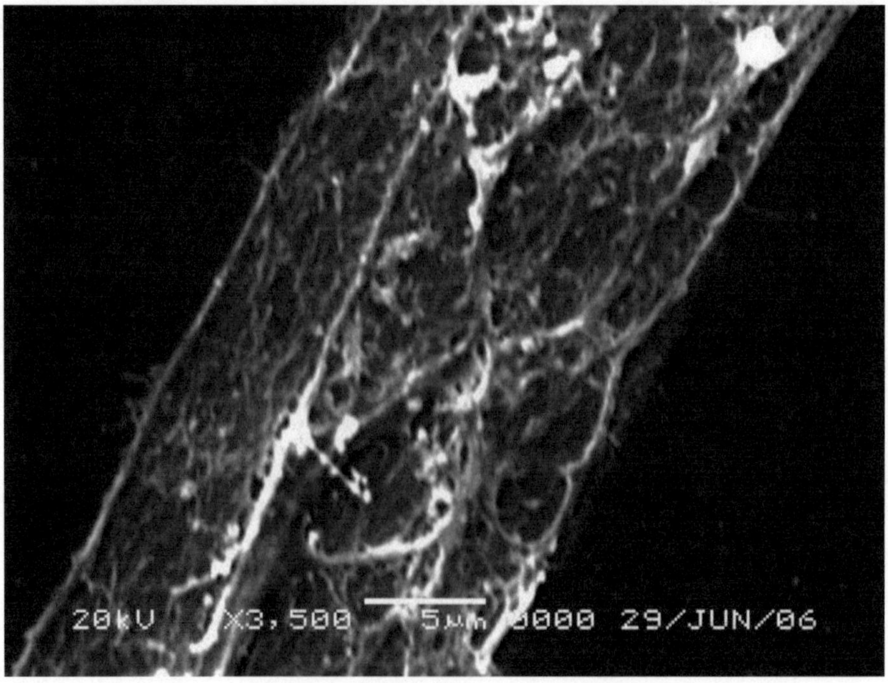

Fig. 16. SEM show that Line of grid were occupied by abound neurites.

4.2.1.2 Examination of synaptic formations by immunocytochemical staining and TEM

To further understand functional activities of neurons on the grid pattern, we observed the microstructure of the neuronal cell body, neuritis extension and processes of the synapse formation by using laser confocal microscope after 7 days in culture. It is found that several neural cell bodies aggregated on the cross point of the grid pattern (Fig.17.b), which neurites extended along the lines clearly by cytoskeletal proteins MAP2 and Synaptophysin double immunocytochemical staining of patterned neurons on PEI(Fig.17.a). Some processes can even gather into a bundle, span the distance between two cross-points and make connection with another neuron. Synaptophysin was a kind of granular protein scattered around the cell bodies and neurites, The neurites from neuron at the cross-point seemed to communicate with those at the lines which express a large number of synaptic vesicle protein(Fig.17.c).

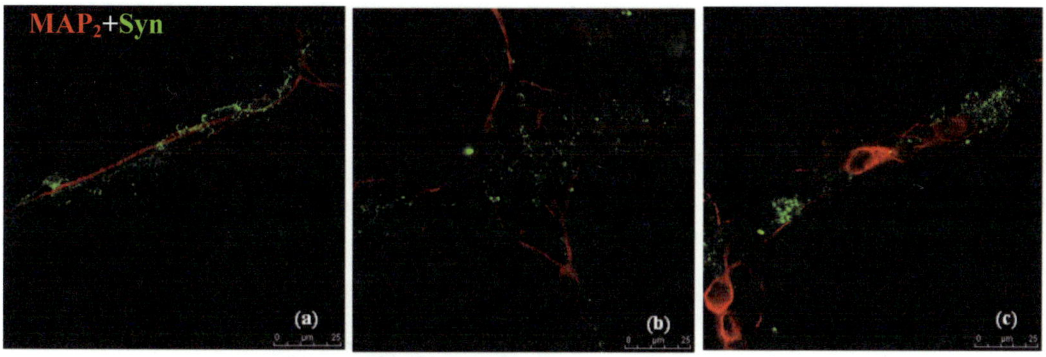

(a) neurites extended along the lines and synaptic vesicle protein around
(b) neural cell bodies at the cross-point and synaptic vesicle protein scattering around them
(c) synapse connection between two neurons at the grid pattern bar=25 μ m

Fig. 17. Immunofluorescent image of anti-MAP2(red fluorescence) labeled neurons and anti-synaptophysin(green fluorescence) labeled synaptic vesicle protein cultured for 7 days on PEI patterns x400

We chose PEI group which can construct more clearly neural network and continued to observe intercellular ultrastructure under transmission electron microscope (Figure18). Two periphery of neurites thicken show high electron density to form a synaptic contact (Red border). Width of synaptic cleft was measured to 30 ~ 50nm. Figure18(a) shows clear synaptic vesicles. Figure18(b) shows the synaptic cleft is relatively narrow and suspected to be electrical synapse structure due to double-membrane structure adjacent closely.

4.2.2 Neural network by microfluidic technique

In this experiment,, we made a big progress on microfluidic technique by re-designing the parameters and enhancing the photoresist coating thickness of the Cr template. After 7 days in culture, poly-l-lysine and laminin+polyethyleneimine were found to be formed more complete and clearer flow patterns by the application of microfluidic technique. On LN group, neurons were easy to cluster into clumps when channel width was 150μm,almost overshadowed flow pattern itself; On PEI group, even though the flow patterns are more complete, the neurites extend short and cannot constitute a connection between some cells

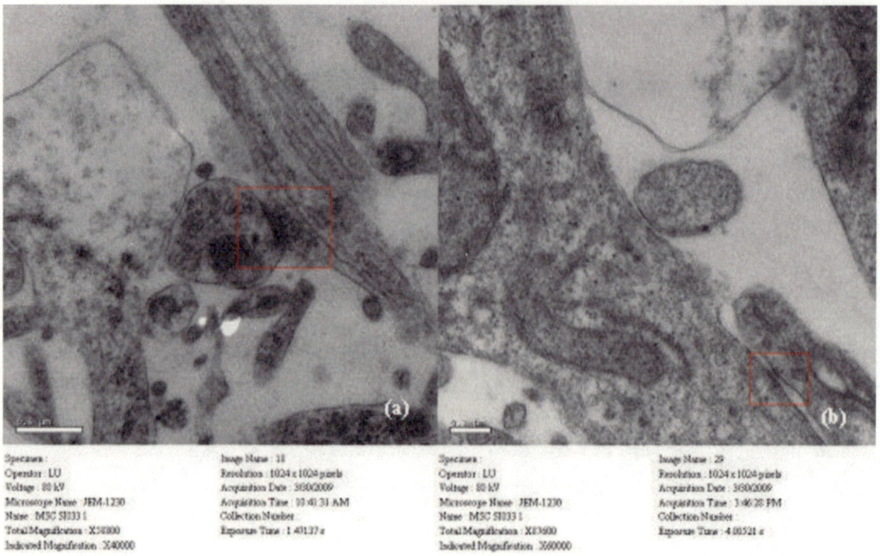

(a)chemical synapse structure x40000 bar=0.5 μ m (b)electrical synapse structure x60000 bar=0.2 μ m

Fig. 18. Under TEM, synapse like structure between neurons from the striatum and the substantial nigra on PEI patterns

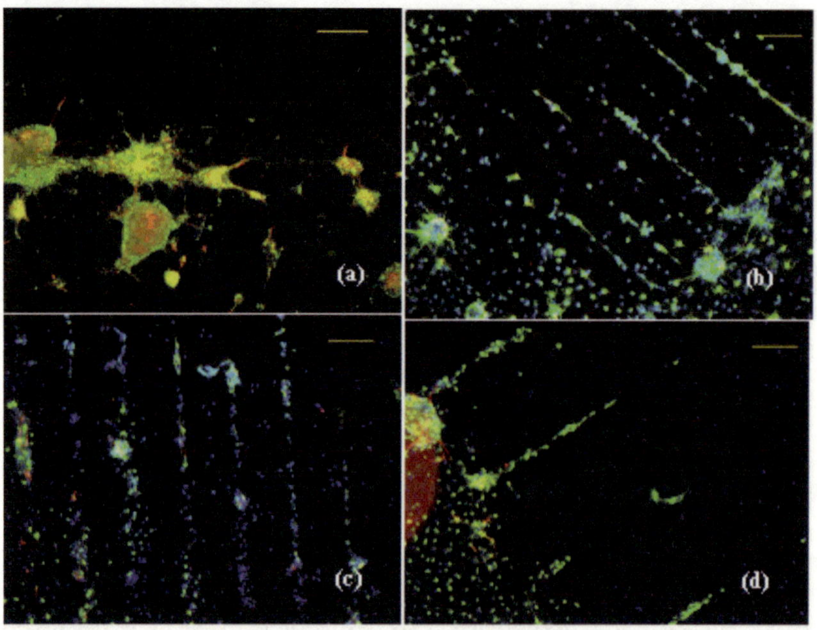

(a) LN group, width of lines 150 μ m, (b) PLL group, width of lines 150 μ m (c) PEI group, width of lines 200 μ m , (d) LN+PEI group, width of lines 300 μ m bar=200 μ m

Fig. 19. Immunofluorescent image of anti-GABA (green fluorescence) +anti-MAP2 (red fluorescence) labelled striatal neurons cultured for 7 days on different substrates by microfluidic technique x100 sufficiently. When the channel spacing of 300μm, LN + PEI group of cells in the liquid injection port at the distribution, flow pattern clear, but the group off between individual cells, so that flow interruption. Identified with immunocytochemical staining, we found that most of the MAP-2 positive neural cells from the striatum cultured on the flow pattern were also labeled with GABA, as shown in Figure 19.

Observing the fine structure of the intercellular on PLL group under scanning electron microscopic in Fig. 20, most of the neural cell bodies were found adhere to the flow pattern and majority of neurites are constrained within the width of the channel to grow following the orientation of flow channels. It is still visible that a few neuronal cell bodies deviate from the flow patterns slightly or adhere on the blank between two lines by local amplification. By observing the cross-linked region off low channel, we can see two endings of neurites seem to have varicose, swelling structure liking synapse.

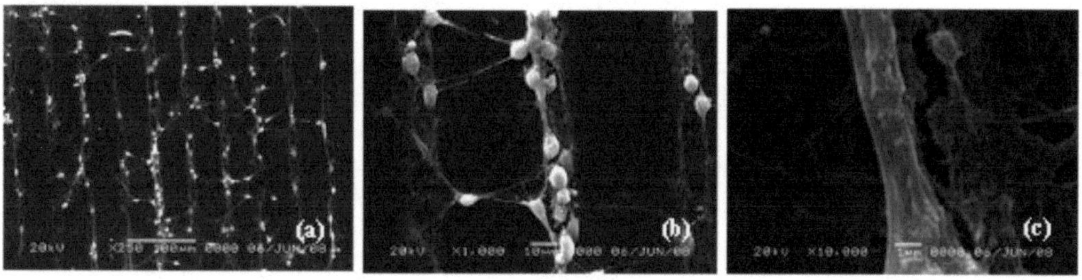

(a)x250 bar=100 μm (b)x1000 bar=10 μm (c)x10000 bar=1 μm

Fig. 20. Neurons from the striatum and the substantial nigra on PLL microfluidic patterns under SEM.

5. Conclusion

In summary, this study based on previous work improve the microfluidic technique, evaluate the influence of two different soft lithography, the micro-contact printing and microfluidic technique on various interface materials for the construction of neural network. Our future work can be divided into two levels. On the one hand continue to look for the intersection of microelectrode array and the micro-fabrication technology, trying to make cell grow in accordance with patterns of MEA electrode nodes. On the other hand, need to further improve the characteristics of electrode materials, enhance biocompatibility under the premise of improving signal-to-noise ratio and without adding resistance, do good to the survival of neurons and neurite extension.

The application of the micro-fabrication on microelectrode array may open up a broader platform of the technique for neurochip research and provide new ideas for the treatment of various injuries in the central nervous system. Ultimately, the combination both can achieve position fixing between neurons and electrodes precisely, to make the system as a real sensor, be able to accept electrical stimulation or chemical stimulation and record their signals, to analysis the transfer process of neural network information, and apply to drug screening of related diseases.

6. References

YN Xia, Whitesides G. M. (1998). Soft lithography. *Angewandte Chemie-International Edition*, Vol.37, No5, pp. 551-575.

Pilnam Kim, Kahp Yang Suh, et al. (2008). Soft Lithography for Microfluidics: a Review. *BIOCHIP* , Vol.2, No1, pp. 1-11,

Tai Hyun Park, Michael L. Shuler. (2003). Integration of Cell Culture and Microfabrication Technology, *Biotechnol. Prog*, Vol.19, pp. 243–253

Y. Nam, J. Chang, and B.C. Wheeler et al. (2004). Patterning to enhance activity of cultured neuronal networks. *Nanobiotechno*, Vol. 151, No. 3, pp. 109-115

Y.Nam, Katherine Musick, Bruce C. Wheeler. (2006)Application of a PDMS micromicrostencil as a replaceable insulator toward a single-use planar microelectrode array, *Biomed Microdevices,* Vol.8, No.4, pp. 375-381

Hiroaki Oka , et al.(1999). A new planar multielectrode array for extracellular recording: application to hippocampal acute slice, *J Neurosci methods*, Vol.93, pp. 61–67

Conrad D. James, et al.(2004). Extracellular recordings from patterned neuronal networks using planar microelectrode arrays, *IEEE Transactions on Biomedical Engineering,* Vol. 51, No. 9, pp. 1640 - 1648

Guangxin Xiang, et al.(2007). Microelectrode array-based system for neuropharmacological applications with cortical neurons cultured in vitro, *Biosensors and Bioelectronics* Vol.22, pp.2478–2484

Falconnet D ,Csucs G, Grandin HM ,et al. (2006). Surface engineering approaches to micropattern surfaces for cell-based assays.*Biomaterials,* Vol.27, No.16, pp. 3044 - 3063

Jackson A , Mavoori J , Fetz EE. (2006). Long-term motor cortex plasticity induced by an electronic neural implant. *Nature,* Vol.444, No.7115, pp. 56 -60

Lemon R. (2007). The Neurochip: Promoting Plasticity with a Neural Implant. *Current Biology,* Vol.17, No.2, pp. 54 -55

Maher M , Pine J , Wright J , et al. (1999). The neurochip: a new multielectrode device for stimulating and recording from cultured neurons. *J Neurosci Methods,* Vol.87, No.1, pp. 45 -56

Jenkner M , Muller B , Fromkerz P. (2001).Interfacing a silicon chip to pairs of snail neurons connected by electrical synapses. *Biol cyber,* Vol.84, No.4, pp. 239 -249

Fan YW , Cui FZ , Chen LN , et al. (2000). Improvement of neural cell adherence to silicon surface by hydroxyl ion implantation. *Surf Coat Tech,* Vol.131, No.1 23, pp. 355 - 359

Fan YW , Cui FZ , Chen LN , et al. (2002). Adhesion of neural cells on silicon wafer with nano-topographic surface. *Applied Surf Sci,* Vol.187, No.3 2 4, pp. 313 - 318

Kumar A , Whitesides GM. (1994). Patterned Condensation Figures as Optical Diffraction Grating. *Science,* Vol.263, No.5143, pp. 60 - 62

Chen, C S. Mrksich, M. Huang, S. Whitesides, G M, et al. (1997). Geometric Control of Cell Life and Death. *Science,* Vol.276, No.5317, pp. 1425–1428

Singhvi, R. Kumar, A.Lopez, G P, et al. (1994). Engineering cell shape and function. *Science,* Vol.264, No.5159, pp. 696–698

Bet, M R. Goissis, G. Vargas, S, et al. (2003). Cell Cell adhesion and cytotoxicity studies over polyanionic collagen surfaces with variable negative charge and wettability adhesion and cytotoxicity studies over polyanionic collagen surfaces with variable negative charge and wettability.*Biomaterials,* Vol.24, No.1, pp. 131–137

Folch, A. Jo, B H. Hurtado, O, et al. (2000). Microfabricated elastomeric stencils for micropatterning cell cultures. *J Biomed Mater Res,* Vol.52, No.2, pp. 346–353

Liu, B F. Ma, J. Xu, QY, et al. (2008). Development of an artificial neuronal network with post-mitotic rat fetal hippocampal cells by polyethylenimine. *Biosens Bioelectron* , Vol.23, No.8, pp. 1221 -1228

3

Toward the Precise Control of Cell Differentiation Processes by Using Micro and Soft Lithography

Yuzo Takayama[1,2], Hiroyuki Moriguchi[3], Kiyoshi Kotani[3], Takafumi Suzuki[1], Kunihiko Mabuchi[1] and Yasuhiko Jimbo[3]

[1]*Graduate School of Information Science and Technology, University of Tokyo*
[2]*Japan Society for the Promotion of Sciences*
[3]*Graduate School of Frontier Sciences, University of Tokyo*
Japan

1. Introduction

Regeneration of the central nervous system (CNS), where proliferative potency is limited, is one of the most important research themes in neuroscience and neuroengineering. Pluripotent stem cell lines have attracted broad attentions as an important model system for regenerating damaged brain. The important key for realizing regenerative medicine using pluripotent stem cells is to induce undifferentiated cell into objective cells with high efficiency and reproducibility. Although the methods to induce embryonic stem (ES) cells into specific neuronal subtype with pharmacological treatment have been proposed, however, the handling is difficult and thus the efficiency rate of objective cells is low (Barberi et al., 2003). The main reason for low differentiation rate of pluripotent stem cells is difficulty in precise control of interaction between pharmacological treatment and cell signalling. To overcome these problems, development of alternative methods for precise control of cell differentiation processes is required.

During cell differentiation, both endogenous factors, such as cell-cell signal transmission, and exogenous factors, such as pharmacological application, play important roles. Micro and soft lithography-based surface modification of culture substrate enabled to control stem-cell-aggregation (EB; Embryoid body) sizes and therefore to increase cell differentiation efficiency through promotion of cell-cell signal interactions (Karp et al., 2007; Wang et al., 2009). Particularly, Park et al. reported that small-size EBs of mouse ES cells (100 ~ 200 μm of diameter) tended to differentiate into ectoderm and large-size EB (500 μm of diameter) into mesoderm (Park et al., 2007). These reports suggested that manipulation of EB size and shape could affect endogenous factors of stem cell EBs and thus was an important approach for precise control of differentiation processes. Another important technique for regulating cell differentiation is applying physical stimulation. It was reported that electrical or magnetic stimulation could induce cellular and molecular responses and affected the gene expressions during differentiation (Kimura et al., 1998; Piacentini et al., 2008). Particularly, Yamada et al. reported that applying electrical stimulation induced mouse ES cells efficiently into ectoderm cells (Yamada et al., 2007). These reports suggested

that applying physical stimulation could affect exogenous factors and also differentiation processes of stem cells. Although the experimental methods to regulate the endogenous or exogenous signalling have been proposed as above, increase in differentiation efficiency is inadequate for regeneration therapy, and combination of the two approaches is not reported.
Based on these problems, we have tried to propose a precise control method for cell differentiation by combining EB size treatment and electrical stimulation technique. To do this, we developed microcavity-array device with embedded electrodes (Takayama et al., 2009). Uniform-sized EBs of P19 cells were prepared and aligned in the electrode substrate, and were stimulated uniformly and simultaneously. However, difficulty in handling EBs and limited number of stimulating EBs hindered further analysis. In this chapter, we describe an alternative method to stimulate more number of EBs based on soft-lithography technique and recent attempt to analyse the effects of EB size treatment and electrical stimulation on the gene expression of P19 cells.

2. Electrical stimulation of EBs with cell-nonadhesive electrode substrates

In this section, we summarize the methods and the results of ensemble electrical stimulation of P19 EBs with the microfabricated electrode substrates, as we reported before (Takayama et al. 2009).
The experimental procedures for fabricating microcavity-array electrode substrate are described as follows. A glass substrate with a transparent conductive layer (ITO; Indium-tin-oxide, Sanyo Vacuum) was cleaned with acetone (Wako) and isopropyl alcohol (IPA, Wako). Then, SU-8 3050 negative photoresist (Microchem) was spin-coated onto an ITO substrate for 60 s at 1000 rpm with a thickness of 100 μm. The coated substrates were baked on a hotplate for 50 min at 95 °C. The substrates were exposed to UV through a custom-made photomask. A reduced projection exposure system (MM-505, Nanometric Technology) was used for fabricating the photomasks. The photomask featured sixteen microcavities with a diameter of 200 or 500 μm aligned in 4 x 4 matrix patterns. After baking on a hotplate for 10 min at 95 °C, the substrates were developed in SU-8 developer (Microchem) and rinsed with IPA. Because SU-8 resist have good insulating properties, the bottom of each microcavity, where the ITO layer was exposed, acted as the stimulation electrode. After fabrication of the microcavity-array pattern, a glass ring of 1 cm height was mounted onto a substrate with a silicone elastomer (KE-103, Shin-etsu Silicones). A stranded wire of cupper was also mounted onto a substrate with an electroconductive paste (Dotite FC-415, Fujikura Kasei). The ITO regions inside a glass ring, except the bottom surfaces of the microcavity, were insulated by coating with a silicone elastomer. The protocols for fabricating the microcavity-array electrode substrate were shown in Fig. 1.
In this research, we used P19 cells as a model system for a stem cell. P19 is a mouse embryonal carcinoma cell line and can differentiate into three germ layers (Bain et al. 1994). For its easy handling and high neuronal differentiation rate, we have used P19 cells as a stem cell model. P19 cells were rountinely cultured and passaged every two days in alpha minimum essential medium (α-MEM, Invitrogen) containing 10 % fetal bovine serum (FBS, HyClone) and 5 - 40 U/ml penicillin-streptomycin (Sigma-Aldrich). To produce uniform size EBs, P19 cells were collected and replated in a 96-well low cell-attachment plate (Sumilon, Sumitomo Bakelite). Due to its spheroid bottom structure, a single EB was formed in each well. The obtained EBs were collected and transferred into

each microcavity of the electrode substrates. To our experience, initial plating of 500 cells resulted in a EB of 200 μm diameter and 4000 cells in a EB of 500 μm diameter at next day of plating. The microcavity-array electrode substrates were previously coated with MPC (2-methacryloyloxyethyl phosphorylcholine) polymer (Lipidure CM5206E, NOF corp.) to prevent nonspecific cell adhesion (Ishihara et al., 1999) and to localize EBs into each microcavity.

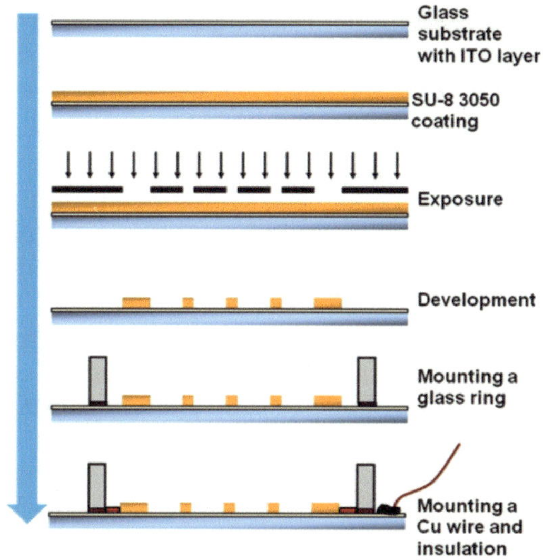

Fig. 1. Schematic of procedures for fabricating the microcavity-array electrode substrate using SU-8 negative photoresist.

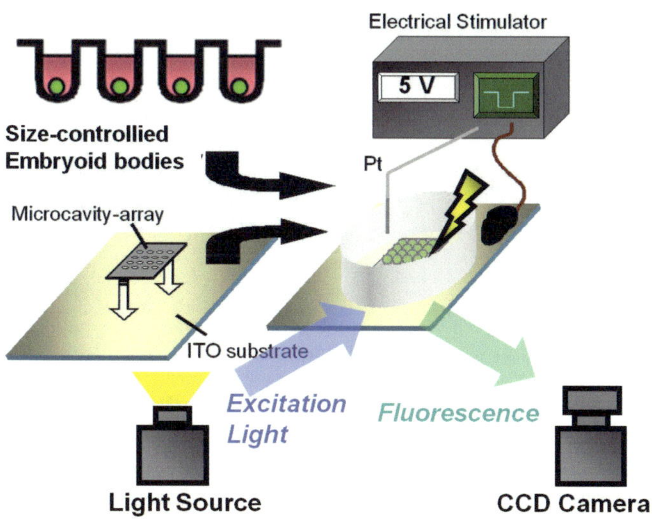

Fig. 2. Schematic diagram of the experimental system. Size-controlled EBs were simultaneously stimulated through the microfabricated ITO substrates.

After insertion of P19 EBs, constant voltage stimulation was applied to the trapped EBs via the bottom surfaces of microcavities with an electrical stimulator (SEN-8203, Nihon Kohden) and an isolator (SS203J, Nihon Kohden). A platinum (Pt) electrode (φ 1 mm) was used as the counter electrode during electrical stimulation. A single negative-first biphasic pulse with intensity of 5 V and duration of 1 ms was used for stimulation. Stimulation-induced responses of EBs were visualized by calcium imaging technique. The EBs were labelled with a calcium indicator Fluo-4AM (Molecular Probes) and fluorescence signals were detected with a cooled CCD camera (C8800-21C, Hamamatsu Photonics) mounted on an inverted microscope (IX-71, Olympus). The frame rate of 0.5 frame/s was used. The recording solution contained 148 mM NaCl, 2.8 mM KCl, 2 mM $CaCl_2$, 1 mM $MgCl_2$, 10 mM HEPES and 10 mM glucose. The overview of the experimental system was shown in Fig. 2.
Figure 3 showed photographs of the fabricated substrate and magnified phase-contrast images of the microcavity-array region and the trapped EBs. These microcavity-array patterns were fabricated with fine reproducibility. The EBs of P19 cells prepared by using spheroid-bottom plates also showed high uniformity in size and morphology. Figure 3-c showed the EBs inserted in the microcavities of 500 μm diameter. The EBs were successfully trapped within each microcavity with 100 μm depth. We confirmed that pre-coating with the MPC polymer inhibited EB adhesion to the device surface over several hours. Thus, suspension culture condition was maintained during electrical stimulation experiments. We also confirmed that pre-coating with the MPC polymer did not affect the electrical properties (impedance, phase) of the device by observing with an LCR meter.

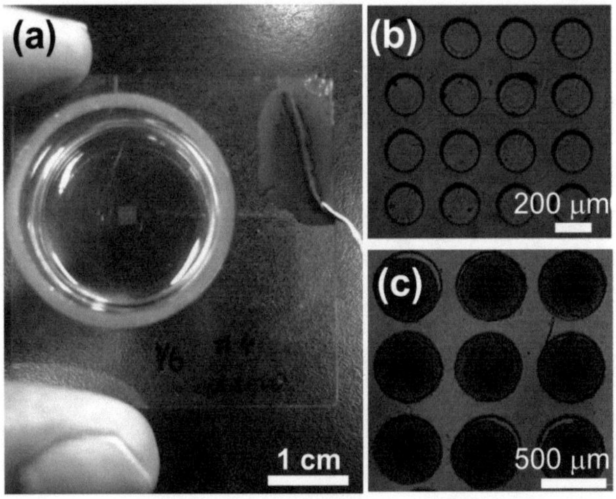

Fig. 3. A microcavity-array dish for ensemble stimulation. (Takayama et al., 2009)

We then applied electrical pulses to the trapped EBs and recorded evoked response by calcium imaging. Figure 4 showed phase-contrast images of EBs in the microcavities and corresponding fluorescence images after applying a single biphasic pulse. The fluorescence images represent the normalized difference ratios in fluorescent intensity between pre- and post-stimulus. Spontaneous calcium transients in cells of P19 EBs were rarely observed before stimulation. By applying electrical stimulation, in contrast, significant elevations of intracellular calcium concentration were observed. For both of microcavity-array patterns, most EBs in the same observing field showed similar responses. In 200 μm diameter EBs, most of cells in an EB showed calcium transients responded to electrical pulse. In 500 μm diameter EBs, cells in outer region of EB primarily showed calcium transients. In all stimulation experiments, we confirmed that similar results were obtained (n=5 for 200 μm diameter patterns and n=3 for 500 μm diameter patterns). The results indicated that the microcavity-array electrode device could stimulate size-controlled EBs simulataneously and uniformally.

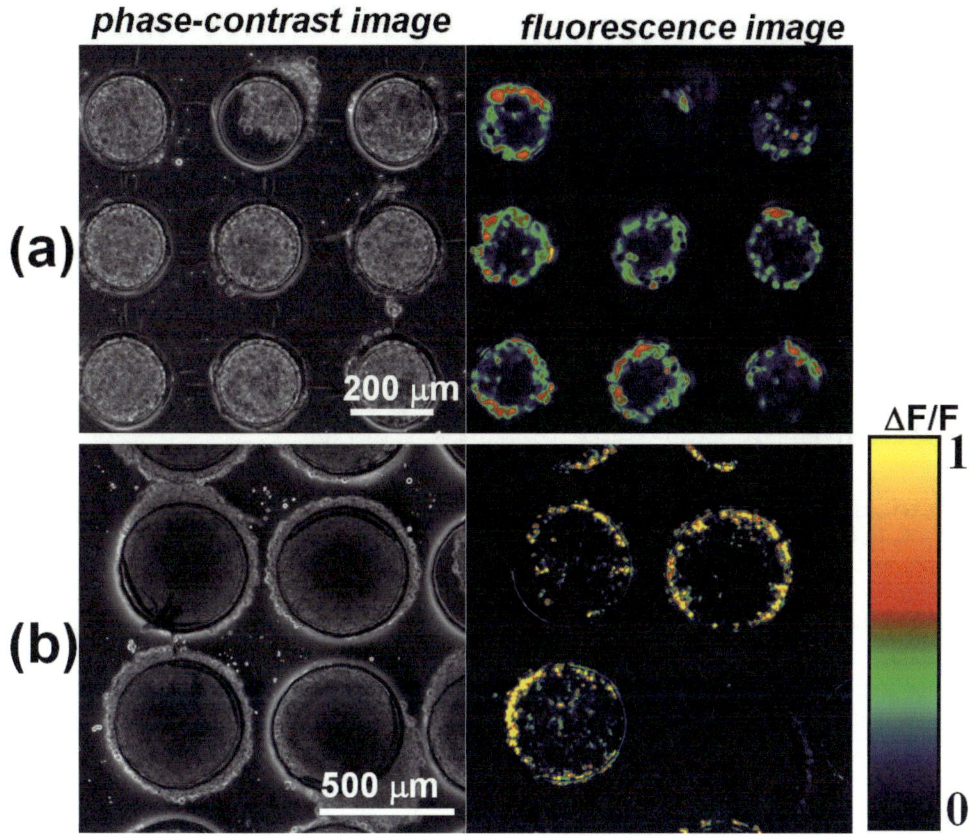

Fig. 4. EBs of P19 cells inserted in the microcavity-array structures (phase-contrast image) and stimulus-evoked intracellular calcium transients (fluorescence image). (Takayama et al., 2009)

In this experiment, by using SU-8 negative thick photoresist, we could fabricate microcavity-array structures onto electrode substrate with enough depth (100 μm) to trap and localize P19 EBs. By using the electrode substrate, we could stimulate P19 EBs simultaneously. Although several studies have carried out electrical stimulation to EBs, such as a field pacing (Sauer et al, 1999 and 2005; Yamada et al., 2007), there have been no studies that observed cell activity in a large number of EBs simultaneously. This may be due to difficulty in observing randomly floating EBs under suspension culture condition. Thus, localizing EBs in specific regions and detecting cell activity of EBs simultaneously by using microcavity-array structures are important bases for further electrical stimulation experiments and analysis.

It is widely recognized that intracellular calcium dynamics play important roles in regulation of cell differentiation processes (Dolmetsch et al., 1998; Yamada et al., 2007). Particularly, Spitzer et al. reported that spatio-temporal patterns of intracellular calcium transients regulated neuronal differentiation processes, such as extension of neurite and selection of neurotransmitter (Spitzer et al., 2004). Thus, inducing and controlling intracellular calcium transients in undifferentiated stem cells by applying electrical stimulation, which parameters are flexibly determined, is promising method to affect cell differentiation processes artificially. Optimizing parameters of electrical stimulation to induce desirable spatio-temporal patterns of intracellular calcium transients will be next step for controlling cell differentiation processes.

There are several problems, however, with the cell-nonadhesive electrode substrate using SU-8 thick photoresist. In the experiments, we manually transferred EBs one by one from 96 well culture plates into each microcavity using micropipettes. Furthermore, due to the cell-nonadhesiveness of the electrode substrate by the MPC polymer coating, mild mechanical perturbations by moving the electrode substrate easily remove EBs from the microcavities. These problems make it difficult to stimulate more than 20 EBs simultaneously with the SU-8-based cell non-adhesive electrode substrates and to perform subsequent gene expression analysis and culture test after stimulation. On the basis of these disadvantages, we attempt to fabricate PDMS (poly(dimethylsiloxane))-based cell-adhesive electrode substrate to stimulate more number of EBs. The experimental procedures and present results for the cell-adhesive electrode substrate are summarized in a next section.

3. Electrical stimulation and gene expression analysis with cell-adhesive microcavity-array electrode substrate

As described in section 2, it was difficult to stimulate a large number of EBs, over a hundred numbers, when using the cell-nonadhesive electrode substrate with SU-8 negative thick photoresist. This was because that the SU-8 structures were strictly integrated with ITO substrate and thus all surface of the electrode substrate became cell-nonadhesive after the MPC polymer coating of microcavity-array region (Fig. 5-a). Although the cell-nonadhesive surface was advantageous to handle EBs, maintain suspension culture conditions and trap EBs into the microcavities temporally, it was disadvantageous to fix EBs to the bottom surfaces of the microcavities not to remove away by mechanical perturbations. Based on these problems, we attempted to propose an alternative method to stimulate a large number of EBs simultaneously. We fabricated cell-adhesive electrode substrate by attaching a PDMS sheet, featuring microcavity-array patterns, onto an ITO substrate (Fig. 5-b). PDMS is well used elastomeric material in cell engineering research because of its capability of easy processing, biocompatibility and cell-nonadhesiveness (Mata et al, 2005). PDMS is also an insulative material (Nam et al., 2006) and thus useful for insulator as the SU-8 structures used in section 2. By using the cell-adhesive electrode substrate with the PDMS sheet and plating P19 cells onto it, we attempted to obtain a large number of EBs and stimulate them. We also attempted to analyse gene expression changes after ensemble electrical stimulation of EBs by polymerase chain reaction (PCR) and electrophoresis Techniques.

The experimental procedures were described as follows. The microcavity-array pattern was fabricated in PDMS using soft lithography technique. First, column-array structures were fabricated as a master mold onto a standard glass substrate using SU-8 photoresist. SU-8 3050 was spin-coated onto a glass substrate (76 × 52 mm, Matsunami Glass) for 60 s at 1000

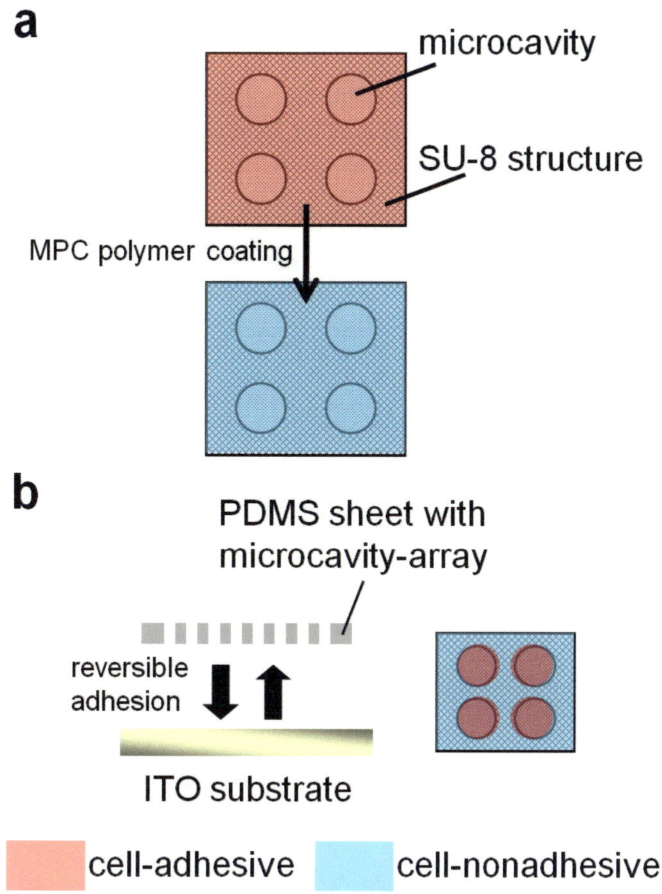

(a) Cell-nonadhesive electrode substrate using SU-8 photoresist.
(b) Cell-adhesive electrode substrate using PDMS elastomer.

Fig. 5. Structures of the electrode sustrates for stimulating EBs

rpm with a thickness of 100 μm. After exposure and development with the photomasks which had inverted patterns as used in section 2, column-array structures were formed on a glass substrate. In this experiment, 289 (17 × 17 matrix pattern) and 64 columns (8 × 8 matrix pattern) were fabricated for 200 and 500 μm diameter microcavities, respectively. Before pouring PDMS, a thin layer of photoresist (OFPR-800, positive photoresist, Tokyo Ohka) with a thickness of 4 μm was formed onto the substrate by spin-coating for 40 s at 2000 rpm and then baking for 30 min at 80 °C. The thin photoresist layer worked as a sacrifice layer to release a cured PDMS sheet from the substrate (Nam et al., 2006). Then, the mixture of PDMS pre-polymer and catalyst (10:1 ratio, Silpot 184, Dow Corning) was spin-coated on the substrate for 80 s at 2500 rpm and 40 s at 2000 rpm, for 200 and 500 μm diameter microcavites, respectively. The PDMS coated substrate was put on a hotplate and then a pre-manufactured thick PDMS annulus (20 mm square and 3 mm thickness) which had a 15 mm diameter hole was immediately attached on the substrate. The PDMS layer and annulus were baked for an hour at 80 °C. After curing, the PDMS sheet was cut along the thick annulus using forceps and then the substrate was soaked in an acetone to solve the photoresist layer. The released PDMS sheet was rinsed with IPA and distilled water. The microcavity-array pattern electrode substrate was formed by attaching the PDMS sheet onto an ITO substrate (50 × 50 mm). The protocols for fabricating the microcavity-array electrode substrate using PDMS were shown in Fig. 6.

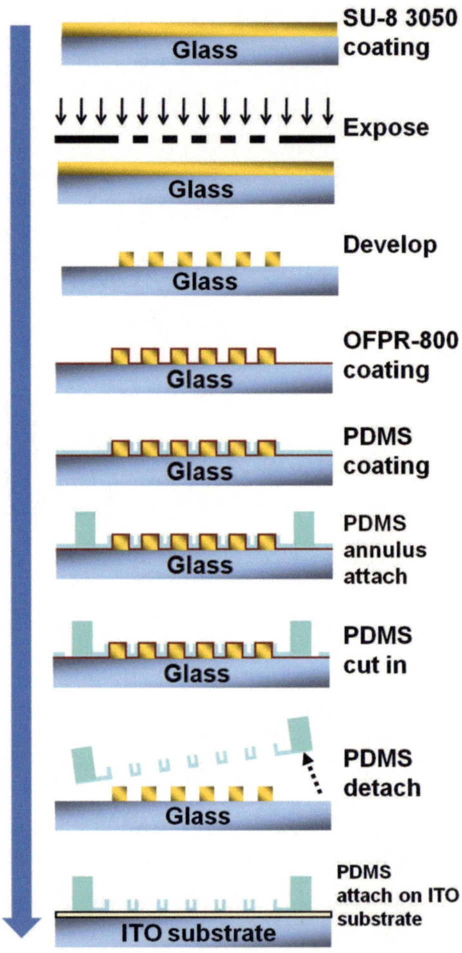

Fig. 6. Schematic of procedures for fabricating the microcavity-array electrode substrate using PDMS elastomer.

Methods for subculture of P19 cells were same as described in section 2. Proliferated P19 cells were collected and then replated onto the microcavity-array electrode substrates. Due to its cell-nonadhesive property of PDMS, P19 cells adhered only to the bottom surfaces of the microcavities. After several hours of plating, non-adhered cells were removed by washing the substrates with culture medium or PBS. P19 cells were proliferated within the microcavities and EBs were self-organizingly formed after 1 day in culture. For electrical

stimulation, a stranded cupper wire was attached to the substrate as described in section 2. Electrical pulses were chronically applied to the EBs with culturing them in an incubator after 1 day of cell plating (EB forming). In this experiment, a single biphasic pulse, same conditions as in section 2, was applied at intervals of 30 min for 2 days.
To analyse the effects of chronic electrical stimulation on cell differentiation processes, reverse transcriptase-PCR (RT-PCR) analysis was carried out after electrical stimulation. First, extraction and purification of RNA were carried out by using a commercial RNA extraction kit (NucleoSpin RNA II kit, Nippon Genetics) from the EBs after applying electrical stimulation for 2 days. RNA of the EBs after induction of neuronal differentiation with retinoic acid was also extracted as the control sample. A 250 ng of total RNA was used for synthesis of first strand complementary DNA (cDNA) by reverse transcription for 80 min at 42 °C using a commercial kit (iScript Select cDNA Synthesis kit, Bio-Rad). Then, amplification of cDNA was carried out with 0.5 μl of synthesized cDNA for 30 s at 94 °C and 1 min at 74 °C, for denaturation and primer extension, respectively. The primer information used in this experiment and primer annealing conditions were summarized in Table 1. The PCR products were analysed using an Experion automated electrophoresis system (Bio-Rad). Temperature controls for reverse transcription and PCR reactions were carried out using a C1000 thermal cycler (Bio-Rad).

Gene Name	Forward Primer	Reverse Primer	Annealing temprature (°C)	Cycles	Size (bp; base pair)	Reference
Oct 3/4	cagaagaggatcaccttggg	gtgagtgatctgctgtaggg	60	27	324	Tang et al., 2002
BMP-4	tgccgcagcttctctgagcc	gctctgccgaggagatcacc	64	26	573	Tang et al., 2002
Mash-1	caagttggtcaacctgggtt	gctcttgttcctctgggcta	56	30	293	Tang et al., 2002
Wnt-1	acgttgctactggcactgac	ccatttgcactctcgcacag	57	30	244	Gao et al, 2001
GAPDH	accacagtccatgccatcac	tccaccaccctgttgctgta	63	25	452	Georgiev et al., 2008

Table 1. Primer sequences and reaction parameters for RT-PCR

The Oct 3/4 gene is thought to play important roles in maintaining undifferentiated conditions and thus is widely used as a marker for undifferentiated stem cells (Ronser et al., 1990). The BMP-4 gene is mainly expressed in ectoderm, meanwhile it inhibit the neuronal differentiation. Thus, an expression level of the BMP-4 is transiently increased and then gradually decreased during neuronal differentiation (Wilson and Hemmti-Brivanlou, 1995). The Mash-1 gene is mainly expressed in neuronal progenitor cells and thus is used as a neuronal marker (Lo et al., 1991). The Wnt-1 gene is also used as a neuronal marker, particularly it inhibit glial differentiation and promote neuronal differentiation (Tang et al., 2002). The GAPDH (glyceraldehyde-3-phosphatedehydrogenase) gene is a widely-used housekeeping gene. These genes are also expressed and regulated in neuronal differentiation processes of P19 cells (Bain et al., 1994).
Figure 7 showed a photograph of the cell-adhesive electrode substrate using a PDMS sheet and a phase-contrast image of the microcavity-array region (diameters of 200 μm in this figure). SEM (scanning electron microscope) images of the SU-8 column-array mold and the corresponding PDMS microcavity-array structures were also shown in Fig. 8. The round-shape microcavity-array patterns as in section 2 were fabricated with fine reproducibility. The depth of the microcavities depended on the height of the column mold (100 μm in this experiment) and could be regulated by changing the experimental parameters.

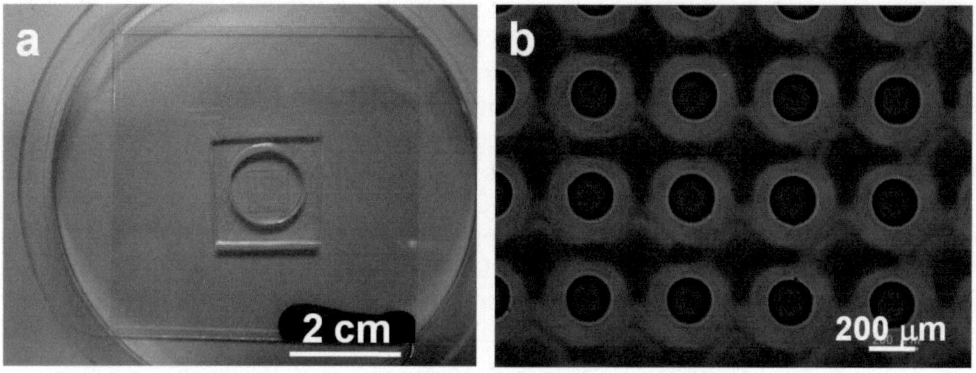

Fig. 7. The cell-adhesive electrode substrate using PDMS (a) the outview of the subsrate. (b) the phase-contrast image of the microcavity-array region.

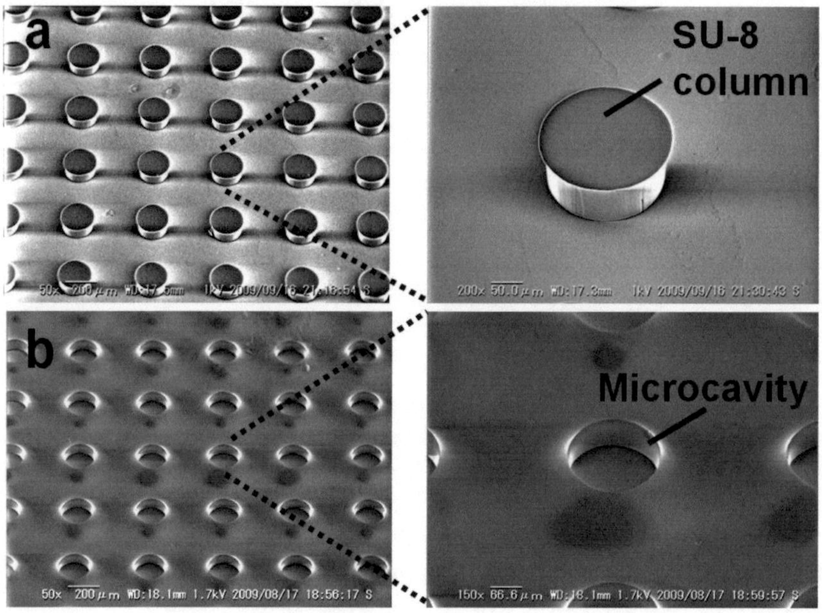

Fig. 8. SEM images of the microcavity-array structures. The diameter of each SU-8 column and microcavity is 200 μm. (a) The SU-8 column-array mold for the microcavites. (b) The corresponding microcavity-array of PDMS.

Then, we plated P19 cells onto the electrode substrate and culture them. The results were shown in Fig. 9. The figure showed the microcavity-array regions of the 200 μm diameter sample (Fig. 9-a) and the 500 μm diameter sample (Fig. 9-b) after 2 days in culture. There were no adhesion and proliferation of P19 cells on the PDMS region. The P19 cells adhered to the bottom surfaces of the microcavities proliferated, aggregated within the microcavities and formed EBs. The characteristic feature of these EBs, different from the EBs in the experiments of section 2, was that they were tightly adhered to the bottom surfaces of the microcavities. Thus, the EBs were not easily remove away by mechanical perturbations. We confirmed that applying electrical pulses evoked intracellular calcium transients in the EBs as those in the experiments of section 2. We concluded that we could obtain a large number of size-controlled P19 EBs and stimulate them simultaneously by using the cell-adhesive microcavity-array electrode substrate.

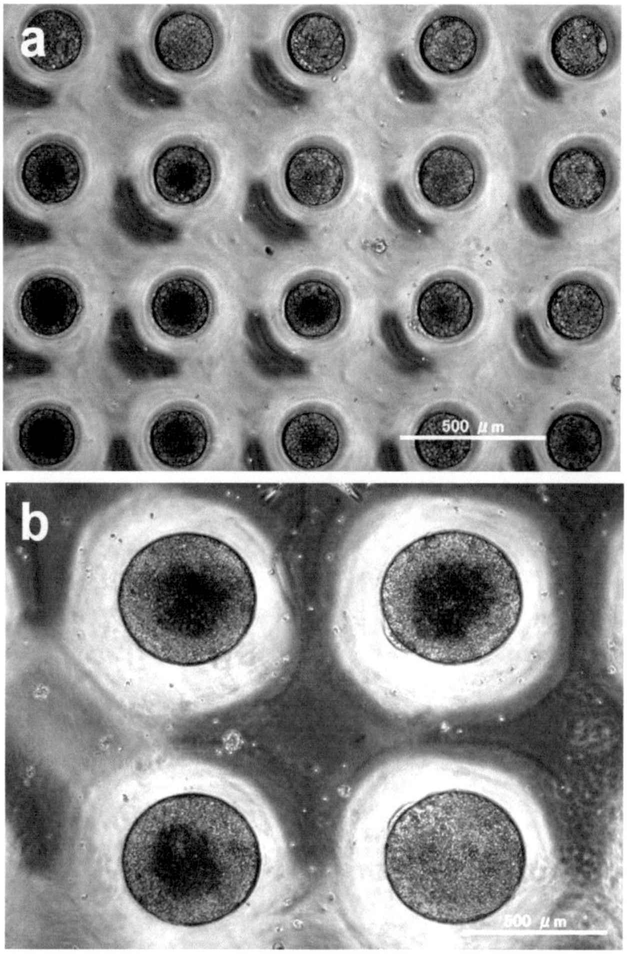

Fig. 9. Self-formed EBs within the microcavity-array substrate using PDMS. (a) the 200 μm diamter pattern sample. (b) the 500 μm diamter pattern sample.

After ensemble electrical stimulation of P19 EBs for 2 days using the cell-adhesive electrode substrate, we attempted to analyse the effects of electrical stimulation on cell differentiation processes of P19 cells by gene expression analysis using a RT-PCR method. Chronic electrical stimulation to P19 EBs was started to be applied after 1 day of cell plating and continued for 2 days with an interval of 30 min. The experimental setup for chronic electrical stimulation of EBs was shown in Fig. 10. A platinum wire was used as a counter electrode as in section 2.

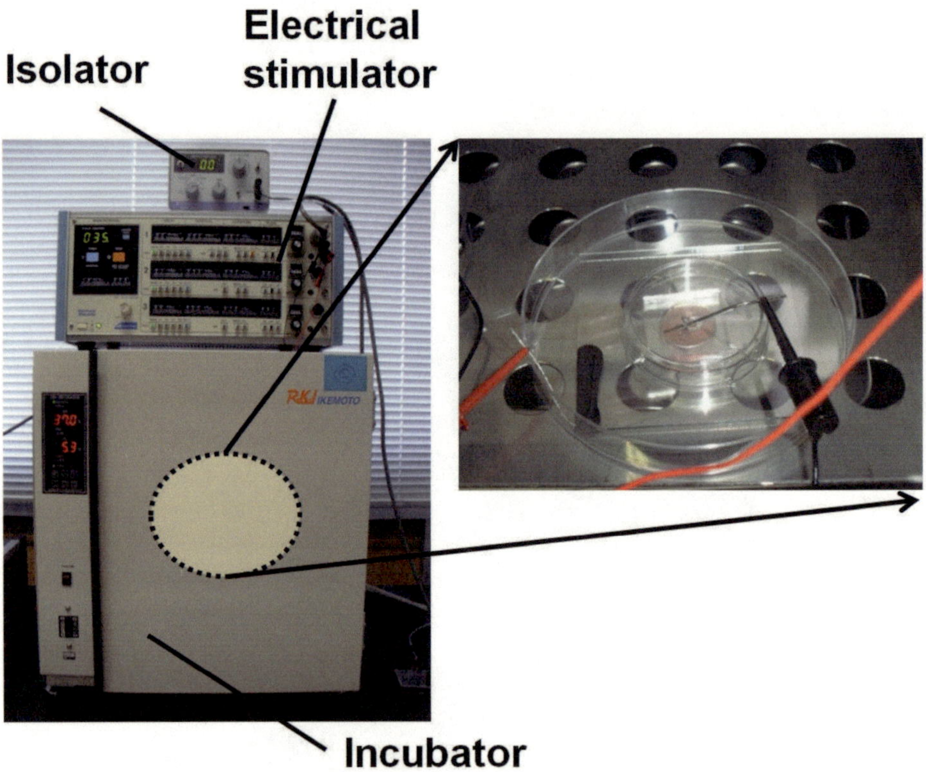

Fig. 10. Chronic electrical stimulation of P19 EBs with the microcavity-array electrode substrate using a PDMS sheet. Electrical pulses were applied for 2 days with an interval of 30 min.

First, we analysed the gene expression of P19 cells induced with retinoic acid under suspension culture conditions for setting a criteria of the gene expression level of neuronal differentiation pathways in P19 cells. To initiate neuronal differentiation, P19 cells were plated into bacteria culture dishes (ϕ 100 mm; Fisherbrand) and were allowed to aggregate with 2×10^6 cells/dish. During the induction period, α-MEM containing 5 % FBS and 1×10^{-6} M all-trans-retinoic acid (Sigma-Aldrich) was used as the culture medium.The expression levels of 5 genes described above were shown in Fig. 11. The figure showed the gene expression patterns under five conditions of P19 cells; undifferentiated cells, EBs under suspension culture conditions with retinoic acid (each of 1 ~ 4 days in culture). The presence of white band indicated the expression of the relevant genes.
The expression of the GAPDH gene was confirmed in all conditions, indicating that PCR reactions in the experiment worked well. In undifferentiated P19 cells, distinct expression of the Oct 3/4 was confirmed, while there were no expression of the BMP-4, Mash-1 and Wnt-1 genes. Following the application of retinoic acid under suspension cultures, the expression of the Oct 3/4 rapidly decreased and disappeared. Alternatively, the BMP-4, Mash-1 and Wnt-1 genes showed elevated expression levels in the EBs. We confirmed the inhibition of undifferentiated associated gene and the expression of neuronal differentiation associated

genes during the differentiation processes of P19 cells following the application of retinoic acid.

Then, the results for gene expression analysis of P19 EBs after application of chronic electrical stimulation using the microcavity-array electrode substrate were shown in Fig. 12. The figure showed the results of two samples for each 200 and 500 μm diameter pattern.

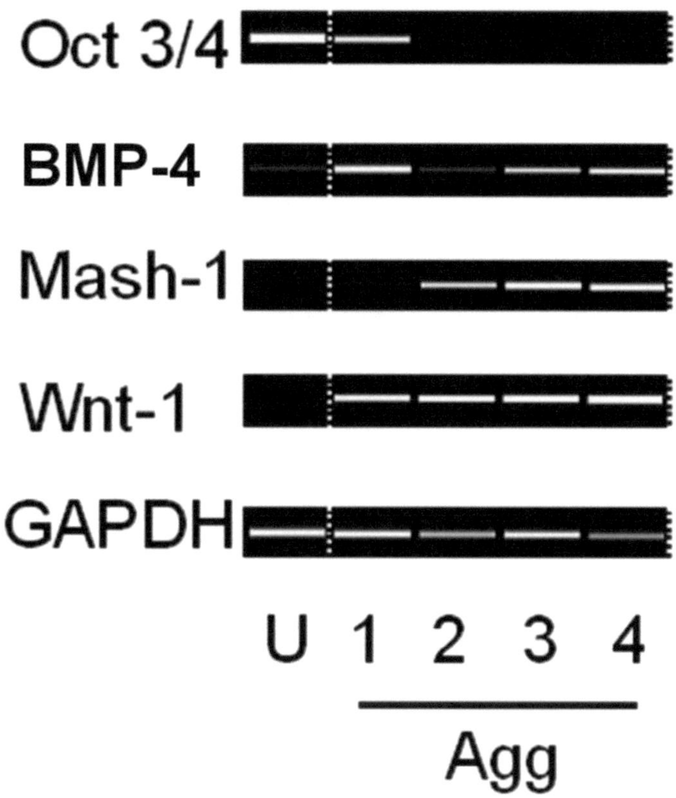

Fig. 11. The gene expression patterns in P19 cells induced by retinoic acid under suspension cultures. U; undifferentiated cells. Agg; EBs in suspension culture. Numbers indicate culture days.

In the four samples used in this experiment, the expression of the Oct 3/4 gene was maintained and there was no distinct expression of the BMP-4, Mash-1 and Wnt-1 genes after electrical stimulation. Thus, in the present situation, we could not induce distinct differentiation processes of P19 cells from its undifferentiated state by electrical stimulation and EBs size treatment.

In this section, we proposed an alternative microcavity-array electrode substrate with PDMS elastomer using micro and soft lithography technique. We could obtain hundreds of size-controlled EBs. The EBs were tightly adhered to the bottom surfaces of the microcavities, thus subsequent electrical stimulation and gene expression analysis could be stably carried out. However, applying electrical stimulation of EBs for 2 days could not induce changes in the gene expression patterns of P19 cells. One of the possible reasons was that an excessive elevation of intracellular calcium concentration activated protease and thus induced cell death (Berridge et al, 1998). Furthermore, while the cell-adhesive surface was advantageous to fix EBs within the microcavities, adhered culture was not suitable condition for induction of cell differentiation (Bain et al., 1994). Optimization of parameters for electrical stimulation and testing other genes will be required.

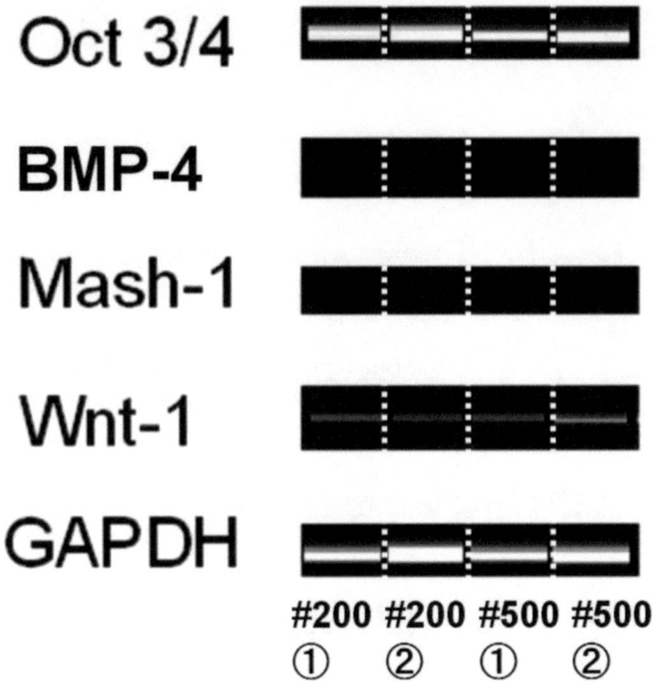

Fig. 12. The gene expression patterns in P19 cells after electrical stimulation for 2 days. #200 indicate the results in 200 μm diamter samples and #500 in 500 μm diamter samples.

4. Conclusion

In this chapter, we proposed the microcavity-array pattern substrate for electrical stimulation of stem cell EBs fabricated by micro and soft lithography technique, aiming at precise control of cell differentiation processes. We could obtain intracellular calcium signals responded to electrical stimulation in EBs simultaneously, indicating that the electrode device would be a promising experimental tool to regulate the cell differentiation processes. We believed that we represented a step on the way to precisely control of cell differentiation processes. Optimization of experimental parameters and improvement of the electrode device structures are current questions.

5. References

Bain, G. ; Ray, W. J. ; Yao, M. & Gottlieb, D. I. (1994). From embryonal carcinoma cells to neurons : The P19 pathway. *Bioessays*, Vol. 16, No.5, 343-348.

Barberi, T. ; Klivenyi, P. ; Calingasan, N. Y. ; Lee, H. ; Kawamata, H. ; Loonam, K. ; Perrier, A. L. ; Bruses, J. ; Rubio, M. E. ; Topf, V. ; Harrison, N. L. ; Beal, M. F. ; Moore, M. A. S. & Studer, L. (2003). Neural subtype specification of fertilization and nuclear transfer embryonic stem cells and application in parkinsonian mice, *Nature Biotechnology*, Vol.21, No.10, 1200-1207.

Berridge, M.J. ; Bootman, M.D. & Lipp, P. (1998). Calcium - a life and death signal, *Nature*, Vol.398, 645-648.

Dolmetch, R.E. ; Xu, K. & Lewis, R.S. (1998). Calcium oscillations increase the efficiency and specificity of gene expression, *Nature*, vol.392, 933-936.

Gao, X. ; Bian, W. ; Yang, J. ; Tang, K. ; Kitani, H. ; Atsumi, T. & Jing, N. (2001). A role of N-cadherin in neuronal differentiation of embryonic carcinoma P19 cells, *Biochemical and Biophysical Research Communications*, Vol. 284, 1098-1103.

Georgiev, D. ; Taniura, H. ; Kambe, Y. ; Takarada, T. & Yoneda, Y. (2008). A critical importance of polyamine site in NMDA receptors for neurite outgrowth and fasciculation at early stage of P19 neuronal differentiation, *Experimental Cell Research*, Vol.314, 2603-2617.

Ishihara, K. ; Ishikawa, E. ; Iwasaki, Y. & Nakabayashi, N. (1999). Inhibition of fibroblast cell adhesion on substrate by coating with 2-methacryloyloxyethyl phosphorylcholine polymers, *Journal of Biomaterials Science, Polymer Edition*, Vol.10, No.10, 1047-1061.

Karp, J. ; Yeh, J. ; Eng, G. ; Fukuda, J. ; Blumling, J. ; Suh, K.-Y. ; Cheng, J. ; Mahdavi, A. ; Borenstein, J. ; Langer, R. & Khademhosseini, A. (2007). Controlling size, shape and homogeneity of embryoid bodies using poly(ethyline glycol) microwells, *Lab on a Chip*, Vol.7, 786-794.

Kimura, K. ; Yanagida, Y. ; Haruyama, T. ; Kobatake, E. & Aizawa, M. (1998). Gene expression in the electrically stimulated differentiation of PC12 cells, *Journal of Biotechnology*, Vol.63, 55-65.

Lo, L. ; Jonson, J.E. ; Wuenschell, C.W. ; Saito, T. & Anderson D.J. (1991). Mammalian *achaete-scute* homolog 1 is transiently expressed by spatially restricted subsets of early neuroepithelial and neural crest cells, *Genes and Development*, Vol.5, 1524-1537.

Mata, A. ; Fleischman, A.J. & Roy, S. (2005). Characterization of polydimethylsiloxane (PDMS) properties for biomedical micro/nanosystems, *Biomedical Microdevices*, Vol. 7, 281-293.

Nam, Y. ; Musick, K. & Wheeler B.C. (2006). Application of a PDMS microstencil as a replaceable insulator toward a single-use planar microelectrode array, *Biomedical Microdevices*, Vol.8, 375-381.

Park, J. ; Cho, C.H. ; Parashurama, N. ; Li, Y. ; Berthiaume, F. ; Toner, M. ; Tilles, A.W. & Yarmush, M.L. (2007). Microfabrication-based modulation of embryonic stem cell differentiation, *Lab on a Chip*, Vol.7, 1018-1028.

Piacentini, R. ; Ripoli, C. ; Mezzogori, D. ; Azzena, G. B. & Grassi, C. (2008). Extremely low-frequency electromagnetic fields promote in vitro neurogenesis via upregulation of Ca_v1-channel activity, *Journal of Cellular Physiology*, Vol.215, 129-139.

Ronser, M.H. ; Vigano, M.A. ; Ozato, K. ; Timmons, P.M. ; Pirie, F. ; Rigby, P.W.J. & Staudt, L.M. (1990). A POU-domain transcription factor in early stem cells and germ cells of the mammalian embryo, *Nature*, Vol.345, 686-692.

Sauer, H. ; Rahimi, G. ; Hescheler, J. & Wartenberg, M. (1999). Effects of electrical fields on cardiomyocyte differentiation of embryonic stem cells, *Journal of Cellular Biochemistry*, Vol.75, 710-723.

Sauer, H. ; Bekhite, M.M. ; Hescheler, J. & Wartenberg, M. (2005). Redox control of angiogenic factors and CD31-positive vessel-like structures in mouse embryonic stem cells after direct current electrical field stimulation, *Experimental Cell Research*, Vol.304, 380-390.

Spitzer, N.C. ; Root, C.M. & Borodinsky, L.N. (2004). Orchestrating neuronal differentiation : patterns of Ca^{2+} spikes specify transmitter choice, *Trends in Neuroscience*, Vol.27, No.7, 415-421.

Takayama, Y. ; Moriguchi, H. ; Saito, A. ; Kotani, K. & Jimbo, Y. (2009). Ensemble stimulation of embryoid bodies using substrate-embedded electrodes, *IEEJ Transactions on Electrical and Electronic Engineering*, Vol.4, 734-735.

Tang, K. ; Yang, J. ; Gao, X. ; Wang, C. ; Liu, L. ; Kitani, H. ; Atsumi, T. & Jing, N. (2002). Wnt-1 promotes neuronal differentiation and inhibits gliogenesis in P19 cells, *Biochemical and Biophysical Research Communications*, Vol. 293, 167-173.

Wang, W ; Itaka, K. ; Ohba, S. ; Nishiyama, N ; Chung, U. ; Yamasaki, Y. & Kataoka, K. (2009) 3D spheroid culture system on micropatterned substrates for improved differentiation efficiency of multipotent mesenchymal stem cells, *Biomaterials*, Vol.30, 2705-2715.

Wilson, P.A. & Hemmati-Brivanlou, A. (1995). Induction of epidermis and inhibition of neural face by Bmp-4, *Nature*, Vol.376, 331-333.

Yamada, M. ; Tanemura, K. ; Okada, S. ; Iwanami, A. ; Nakamura, M. ; Mizuno, H. ; Ozawa, M. ; Ohyama-Goto, R. ; Kitamura, N. ; Kawano, M. ; Tan-Takeuchi, K. ; Ohtsuka, C. ; Miyawaki, A. ; Takashima, A. ; Ogawa, M. ; Toyama, Y. ; Okano, H. & Kondo, T. (2007). Electrical stimulation modulated fate determination of differentiating embryonic stem cells, *Stem Cells*, Vol.25, 562-570.

Part 3

Self-Assembled Colloidal Substrates for Nanofabrication

4

Fabrication of Surfaces with Bimodal Roughness Through Polyelectrolyte/Colloid Assembly

Christine C. Dupont-Gillain*, Cristèle J. Nonckreman, Yasmine Adriaensen and Paul G. Rouxhet
Institute of Condensed Matter and Nanosciences - Bio & Soft Matter, Université Catholique de Louvain
Belgium

1. Introduction

From bioengineering to optics and electronics, a great deal of work has been conducted on the development of new materials with structured surfaces. A large range of methods has been used, such as plasma etching, electron beam and colloidal lithography (Denis et al., 2002), electrical deposition (Yang et al., 2009), phase separation (Dekeyser et al., 2004) and polyelectrolyte assembly in order to produce structured surfaces (Agheli et al., 2006). The combination of different methods is also more and more explored. Schaak et al. (2004) described a simple approach to achieve colloidal assembly on a patterned template obtained by lithography.

Densely packed layers of colloidal particles can be produced by lifting a substrate vertically from a suspension (Fustin et al., 2004; Li J. et al., 2007) or by spin coating (Yang et al., 2009). Colloidal lithography utilizes the ability of particles to adhere on oppositely charged surfaces (Johnson & Lenhoff, 1996; Hanarp et al., 2001, 2003)), possibly using surface modification by inorganic or organic polyelectrolytes. The surface coverage is influenced by several factors: ionic strength, particle size and time. The adhesion of microbial cells on various substrates was also achieved by surface treatments with inorganic or organic polycations (Changui et al., 1987; Van haecht et al., 1985) or with positively charged colloidal particles (Boonaert et al., 2002). A review on colloidal lithography and biological applications was published recently (Wood, 2007).

Adsorption of polyelectrolytes is influenced by ionic strength, pH and the polyelectrolyte characteristics (molecular mass, charge density) (Lindquist & Stratton, 1976; Davies et al., 1989; Choi & Rubner, 2005). At low ionic strength, highly charged polyelectrolytes adopt extended conformations and are fairly rigid due to the strong repulsion between charged units. The maximum adsorbed amount and the adsorbed layer thickness do not vary markedly according to molecular weight. As the salt concentration is increased and the electrostatic intrachain repulsion is decreased, the polyelectrolyte becomes more coiled. In this case, the maximum amount adsorbed (expressed in mass) increases as a function of molecular weight (Roberts, 1996; Lafuma, 1996; Claesson et al., 2005; Boonaert et al., 1999).

Build up of polyelectrolyte films may be achieved using layer-by-layer assembly through alternating adsorption of oppositely charged polyelectrolytes (Decher & Hong, 1991).

Polyethyleneimine (PEI), polyallylamine (PAH), poly-l-lysine (PLL) and polydiallyldimethylammonium chloride (PDDA) are common polycations used for multiple layer formation with polyanions such as polystyrene sulfonate (PSS) (Bertrand et al., 2000). Layer-by-layer assembly of polyelectrolytes has been combined with the use of colloidal particles. For instance, a film was made on silicon wafers precoated with thermally evaporated titanium, by adsorption of PDDA, followed by adsorption of PSS, followed by treatment with an aluminium chloride hydroxide solution (Hanarp et al., 2001, 2003). In another study, chemically patterned surfaces made by self-assembled monolayers (SAMs) were covered with a polyelectrolyte multilayer film, before adhesion of SiO_2 silica particles or functionalized polystyrene latex particles (Chen et al., 2000). While a polyelectrolyte layer may provide a strong bond to colloidal particles, the drying process applied after particle adhesion may be crucial to obtain a regular and homogeneous monolayer (Hanarp et al., 2003).
Structured hydrophobic surfaces have gained increasing interest because the roughness amplifies the hydrophobicity (Wenzel, 1936). This is exemplified by the Lotus effect, in which a dual size roughness seems to be important (Barthlott & Neinhuis, 1997; Patankar, 2004). Raspberry-like surface morphologies were created in different ways : styrene polymerization (Perro et al., 2006; Reculusa et al., 2002) or gold sputtering (Xiu et al., 2006) on silica particles , controlled aggregation of different surface-functionalized silica particles (Ming et al., 2005) or direct electrochemical synthesis of gold microaggregates (Li Z. et al., 2007) and immobilization on a specific substrate.
In this paper, we prepare surfaces covered with a homogeneous monolayer of colloidal particles, using adhesion of negatively charged polystyrene latex beads on a polycation-precoated glass substrate. The method is then extended to prepare surfaces presenting a bimodal roughness, by using latex particles of different sizes. The influence of substrate surface roughness on the behavior of mammalian cells has been of great concern in the last years (Nonckreman et al., 2010). Therefore, the stability of fabricated samples is tested in phosphate buffer saline (PBS), which simulates the pH and ionic strength of biological fluids. Note that here, the term "colloid" is used with the restrictive sense of lyophobic colloidal particle, and thus distinguished from polyelectrolytes.

2. Materials and methods

2.1 Materials

Glass substrates were microscope coverslips (12 mm diameter, Menzel-Gläzer, Germany). Polyethyleneimine solution (PEI, M_w = 750 000), polyallylamine hydrochloride (PAH, M_w~70 000), poly-l-lysine hydrobromide (PLL, M_w = 70 000 to 150 000), poly(diallyldimethylammonium chloride) solution (lPDDA, M_w < 100 000; hPPDA, M_w = 400 000 to 500 000) were purchased from Sigma-Aldrich (St. Louis, MO, USA). Negatively charged polystyrene particles with two different sizes (sulfate latex, 65±6nm, 470±12nm) were purchased from Interfacial Dynamic Corporation (IDC, Tualatin, OR, USA). The composition of the phosphate buffer saline solution (PBS) was as follows: 137 mM NaCl, 6.4 mM KH_2PO_4, 2.68 mM KCl, 8 mM Na_2HPO_4; adjusted to pH 7.4 with 1M NaOH.

2.2 Sample preparation

Figure 1 presents the different steps of sample preparation, from the glass substrate to surfaces with bimodal roughness.

2.2.1 Glass substrate conditioning

The glass coverslips were cleaned by overnight immersion in sulfochromic solution and rinsed with water, prior to polycation adsorption (Figure 1a). The pH and ionic strength (I) of the polycation solution were adjusted with NaOH and HCl, and NaCl, respectively. The polycation solution (1 ml) was poured into the wells of a tissue culture plate (Falcon, Becton Dickinson, Belgium, Ref. 353226), where the glass coverslips had been placed earlier, and was left in contact with the substrate for at least 2 h. Unless stated otherwise, the polycation solution was 10^{-5} M at pH 11 and I 10^{-2} M. The samples were rinsed by 6 successive dilutions to avoid exposure to air. Each rinsing step was performed by adding 2 ml of deionized water (produced by a Milli-Q plus system from Millipore, Molsheim, France), stirring gently, and removing 2 ml of liquid. They were then dried under a gentle nitrogen flow (Figure 1b).

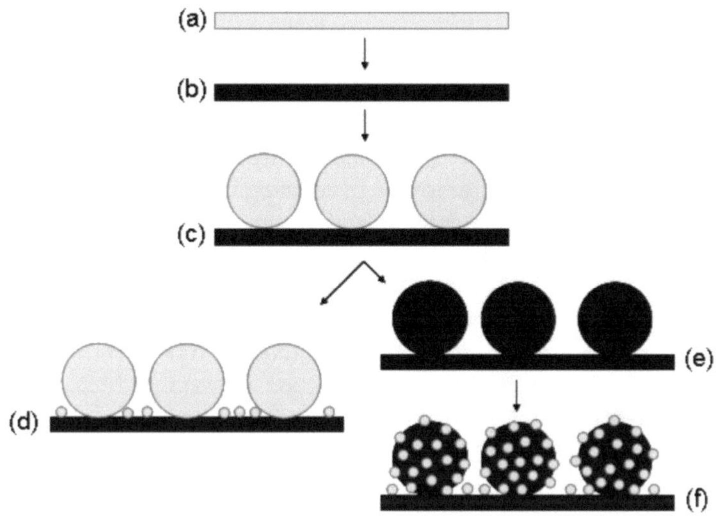

Fig. 1. Schematic representation of the steps used for sample fabrication. Glass substrate (a), polycation-conditioned glass substrate (b), substrate with a layer of adhering particles (c), surface with bimodal I roughness (d), substrate with a layer of adhering particles conditioned with the polycation (e), surface with bimodal II roughness (f).

2.2.2 Adhesion of a single colloid layer

Polycation-conditioned substrates were placed horizontally into wells of another culture plate. 1 ml of a suspension of the desired concentration, pH and ionic strength was poured into the wells and left in contact with the substrate for at least 3 h (Figure 1c). Unless stated otherwise, the procedure used for rinsing and drying the samples was performed as follows: rinsing 3 times with water and 3 times with isopropanol and drying overnight in air. The samples were rinsed by successive dilutions as detailed in the preceding paragraph. After the last rinsing, as much as possible of the solution was removed with a pipette.

2.2.3 Surfaces with bimodal I roughness

1 ml of a 0.1 % suspension of 470 nm particles was poured into the wells containing glass substrates conditioned with PAH (10^{-5} M, pH=11, I=10^{-2} M) (Figure 1b), placed in horizontal

position, and left in contact with the substrate for at least 3 h (Figure 1c). A suspension of 65 nm particles was then added to obtain a final concentration of 0.1 %, and the system was gently stirred and left resting for 1 more hour (Figure 1d). The samples were rinsed and dried using the same procedure as described for adhesion of a single colloid layer.

2.2.4 Surfaces with bimodal II roughness

Substrates with adhering 470 nm particles (Figure 1c), rinsed and dried, were again conditioned with PAH (10^{-5} M, pH=7, I=10^{-3} M) (Figure 1e), rinsed with water then with isopropanol and dried, as described for the adhesion of a single colloid layer. These samples were then placed horizontally in wells of another culture plate. 1ml of a 0.1% suspension of 65 nm particles was poured into wells and left in contact with the substrate for at least 2 h (Figure 1f). The samples were rinsed and dried using the same procedure as described for adhesion of a single colloid layer.

2.3 Surface characterization

The samples were examined by scanning electron microscopy (SEM) using a high resolution Digital Scanning microscope 982 Gemini (Leo Electron Microscopy, UK) operating at 1 or 2 kV, without any metal coating. SEM images of substrates covered with a monolayer of adhering particles were analyzed after image processing with Adobe Photoshop (version 8). In this way, the white rings defining latex particles were extracted and intensified in order to draw circles for digital processing. The images were binarized and the circles were filled before measurements (Visilog, version 6). A minimum of 500 particles was counted on more than 4 images with 10 000x magnification. The degree of particle coverage was measured as the area of particles divided by the total area. Manual counting was also performed on at least 3 images with 10 000x or 20 000x magnification of each sample. Regarding surfaces with bimodal roughness, no image processing was found suitable for extracting colloidal particles from the background. Manual counting of the 470 nm particles only was performed on at least 4 images.

The surface chemical analysis by X-ray photoelectron spectroscopy (XPS) was performed using a Kratos Axis Ultra Spectrometer (Kratos Analytical, UK) equipped with a monochromatized aluminium X-ray source (powered at 10 mA and 15 kV) and eight channeltrons detector. Charge stabilization was achieved using an electron source (filament current set at 1.9 A, bias -1.1 eV) mounted coaxial to the lens column and a charge balance plate (voltage set at -2.3 V). The rectangular analyzed area was about 0.7 mm × 0.3 mm. For recording individual peaks, the pass energy was set at 40 eV. In these conditions, the full width at half maximum (FWHM) of the Ag $3d_{5/2}$ peak was about 0.9 eV. The pressure in the analysis chamber was around 10^{-6} Pa. The following sequence of spectra was recorded: survey spectrum, C 1s, O 1s, N 1s, Si 2p, Cl 2p, B 1s, Na 1s, Zn 2p, Ti 2p, Al 2p, S 2p and C 1s again to check the stability of charge compensation and the absence of sample degradation as a function of time. The binding energy scale was set by fixing the C 1s component due to carbon only bound to carbon and hydrogen at 284.8 eV. The data analysis was performed with the CasaXPS program (Casa Software, UK). Molar concentration ratios were calculated using peak areas normalized on the basis of the acquisition parameters, sensitivity factors, and transmission function provided by the manufacturer. Angle-resolved XPS analyses were performed by using angles between the normal to the sample surface and the direction of photoelectron collection, φ, equal to 0°, 45° and 60°.

The streaming potential was determined at 20°C with an instrument purchased from the Department of Physical and Colloid Chemistry (Agricultural University, Wageningen, The Netherlands) (Elgersma et al., 1992) using glass microscope slides from Marienfeld (7.6 x 2.6 cm^2; Lauda-Königshofen, Germany). Two plates of the samples to characterize (clean glass, PAH-conditioned glass) were assembled to make a 100 μm-thick chamber. A solution of 0.01M KNO_3 was forced through these two plates. The difference of potential measured at the entrance and the exit of the chamber was used to compute to the zeta potential of the substrate (Rouxhet et al., 1993).

3. Results and discussion

3.1 Surface conditioning with PAH

The choice of pH = 11 for PAH adsorption was inspired by the following considerations. The pK_a of ethyl ammonium is 10.6. The apparent pK_a of the PAH used here is 8.7 in water and 9.3 in 0.5 M NaCl (Petrov et al., 2003; Choi and Rubner, 2005); however it is expected to be appreciably higher after adsorption by a negatively charged surface (Tagliazucchi et al., 2008). Conditions of low degree of protonation were chosen in order to allow adsorption of a thick layer. The adsorbed amount was indeed reported to be higher for polymers with low and intermediate cationicities, such as PAH or PLL compared to PEI, and to increase with the molecular mass (Roberts, 1996; Lafuma, 1996), owing to a more coiled conformation. Highly charged polycations, such as PEI, form flat adsorbed layers at low ionic strength (Claesson et al., 2005) and the adsorbed amount increases with pH (Meszaros, 2004).

The surface chemical composition of glass and glass conditioned with PAH (3 independent sets of results) is presented in Table 1. Non-conditioned glass showed the expected organic contamination and a low concentration of nitrogen. In addition, low concentrations of potassium (1-2 %), boron (2-3 %), sodium (1-1.5 %), and traces (< 1 %) of zinc, titanium and aluminium were found. As the analyzed depth decreased (increase of φ), the C/Si concentration ratio increased as expected, but N/Si did not vary significantly and N/C decreased. This indicates that nitrogen is associated with the glass matrix or the glass surface and not with the organic contaminants.

After PAH conditioning, polycation adsorption was evidenced by the increase of carbon and nitrogen concentrations and the decrease of silicon and oxygen concentrations. The N_{1s} peak showed components at 401.5 eV and 399.3 eV due to protonated and non protonated nitrogen, respectively. As the analyzed depth decreased (higher φ), the apparent concentration of carbon and nitrogen increased as expected, the N/C ratio remained constant, but the proportion of protonated nitrogen decreased appreciably.

The N/C ratio of about 0.14 must be compared with the value of 0.33 expected for pure PAH. This difference may not be due to an orientation of the polymer segments at the surface, given the structure of the repeat unit [$-CH_2-CH(-NH-CH_3)-$] and the analyzed depth. It is attributed to the simultaneous presence at the surface of PAH and organic contaminants (Caillou et al., 2008). As the N/C ratio does not vary with φ, the adsorbed organic layer appears to be a mere mixture, with no preferential accumulation of a component at the outermost surface. In an alternative way, the elemental composition of the adsorbed layer may be estimated by the difference of nitrogen concentration before and after PAH conditioning ratioed to the total carbon concentration, which provides a N/C ratio of 0.11. The adventitious contaminants observed on silica, which showed the same C1s peak shape as that observed here on glass, were modeled by the generic formula $C_{15}H_{28}O_4$

(Gerin et al., 1995). According to a N/C ratio of 0.11, the adsorbed layer obtained after PAH conditioning would be 33 wt% PAH (density 1.04 g/m³) and 67 wt% $C_{15}H_{28}O_4$ (density of 0.9 g/cm³).

Substrate	φ	Concentration (Mole fraction, in %) C	O	Si	N	S	Concentration ratios C/Si	N/C	N^+/N_{tot}
Glass coverslips									
sample 1	0°	12.6	57.2	22.8	0.6	-[a]	0.55	0.05	0.47
	45°	15.6	58.4	19.7	0.4	-[a]	0.79	0.03	0.37
	60°	20.5	56.5	18.0	0.4	-[a]	1.14	0.02	0.33
sample 2	0°	10.1	59.4	24.9	0.9	-[a]	0.41	0.09	-[b]
	60°	16.5	59.6	20.1	0.8	-[a]	0.82	0.05	-[b]
sample 3	0°	11.9	57.7	23.6	0.9	-[a]	0.50	0.07	0.46
PAH pre-coated glass									
sample 4	0°	24.1	46.6	18.6	3.5	-[a]	1.30	0.15	0.31
	45°	31.9	43.3	14.3	4.8	-[a]	2.23	0.15	0.18
	60°	42.5	36.9	11.3	5.6	-[a]	3.76	0.13	0.12
sample 5	0°	22.8	48.5	19.5	3.4	-[a]	1.17	0.15	-[b]
	60°	37.6	40.9	12.5	5.2	-[a]	3.01	0.14	-[b]
sample 6	0°	30.4	42.5	18.6	4.5	-[a]	1.63	0.15	0.49
PAH pre-coated glass with adherent colloidal particles									
sample 7 (470 nm)	0°	64.1	21.8	8.1	1.6	0.3	7.91	0.02	0.38
sample 8 (470 nm)	0°	41.1	38.0	13.9	2.7	0.2	2.96	0.06	-[b]
sample 9 (65 nm)	0°	52.2	30.4	11.3	2.7	0.4	4.62	0.05	-[b]

Table 1. Surface chemical composition determined by XPS (mole fraction with respect to all elements excluding hydrogen) at different photoelectron collection angles φ.

In order to evaluate the thickness of the organic adlayer, the experimental C/Si ratios were compared with ratios computed by considering a layer of constant thickness (t) on top of glass and using equation 1:

$$\frac{C}{Si} = \frac{i_{Si}}{i_N}\frac{\sigma_C}{\sigma_{Si}}\frac{\lambda_C^{ad}\, C_C^{ad}\left[1-\exp\left(-t/\lambda_C^{ad}\cos\varphi\right)\right]}{\left[\lambda_{Si}^{su}\, C_{Si}^{su}\exp\left(-t/\lambda_{Si}^{ad}\cos\varphi\right)\right]} \qquad (1)$$

where i_{Si} and i_N are the relative sensitivity factors provided by the manufacturer for Si (0.328) and C (0.278); σ_{Si} and σ_N are the photoionization cross sections for Si (0.817) and C

(1.000). The superscripts ad and su designate the adsorbed overlayer and the substrate, respectively. The concentrations C are: C_C^{ad} = 49.6 mmol/cm³ for contamination layer deposited on glass; C_C^{ad} = 46.9 mmol/cm³ for the layer made by a mixture of 67 % contaminents and 33 % PAH (N/C = 0.11); C_{Si}^{su} = 39 mmol/cm³ for Si in glass (considering that Si concentration in glass is about 90 % of that in pure SiO_2; density of 2.6 g/cm³). The electron inelastic mean free paths were calculated according to Tanuma et al. (1997) with an energy gap of 7 eV: λ_C^{ad} = 3.9 nm, λ_{Si}^{ad} = 3.6 nm, λ_C^{ad} = 4.4 nm. Figure 2 presents plots of C/Si molar ratios computed using different values of t, as a function of φ. The curves are only slightly different if a N/C ratio of 0.14 is considered. Comparison with the experimental data indicates that the organic layer adsorbed after conditioning is about 3 nm, which is about 3 times thicker than the contamination layer. This does not fit closely the apparent proportions of contaminants and PAH in a homogeneous adlayer. Such approach cannot be exploited further owing to the simplicity of the model (homogeneous, flat and compact adlayer) and limited data accuracy. However it indicates that the organic adlayer obtained after conditioning has a thickness of about 3 nm and that PAH coexists with contaminants. The latter may adsorb from the surrounding atmosphere between sulfochromic cleaning and PAH conditioning or after conditioning.

The degree of protonation of PAH in a solution at pH 11 is less than 10 %. The XPS spectra recorded after PAH conditioning (Table 1) gave values of 31 and 49 % for the degree of nitrogen protonation at a photoelectron collection angle φ = 0°. This is in agreement with expectations regarding the effect of local potential on protonation and with data indicating that the apparent pK_a of PAH may be appreciably increased on a negatively charged surface (Tagliazucchi et al., 2008).

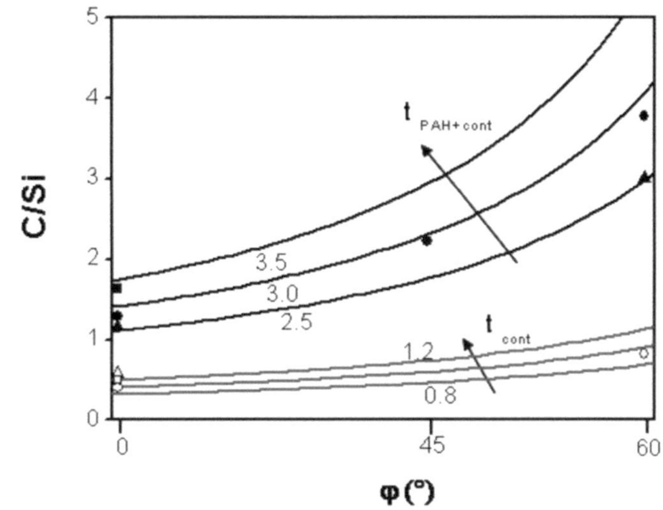

Fig. 2. Plot of computed C/Si molar concentration ratios vs the angle φ of photoelectron collection, for non-conditioned glass with a 0.8, 1.0, 1.2 nm thick (t_{cont}) contamination layer, and for PAH-conditioned glass with a 2.5, 3.0 and 3.5 nm thick ($t_{PAH+cont}$) adlayer made of PAH and contamination. Experimental data for glass (open symbols) and for PAH-conditioned glass (dark symbols). Circles: samples 1 and 4, triangles: samples 2 and 5, squares: samples 3 and 6 (see Table 1).

Figure 3 presents the variation of the zeta potential of glass and of PAH-conditioned glass as a function of pH (in 0.01 M KNO_3). As expected, the glass substrate was found to be negatively charged above pH 2. After polycation adsorption, the zeta potential became less negative, but did not reach positive values. It appears (Figure 3) that adsorbed PAH is shielding the negative charge of glass surface but is not reversing the surface charge. This is in contrast with data reported for substrates such as mica and titanium oxide (Adamczyk et al., 2006, 2007) and is attributed to the fact that here the samples were dried between adsorption and streaming potential measurement, which provoked irreversible shrinkage of the adsorbed layer. It may be related to the decrease of the degree of nitrogen protonation as φ increased (Table 1), revealing that it is much higher for segments in close contact with the glass surface. Note that drying after adsorption was a choice made in the context of a fabrication process.

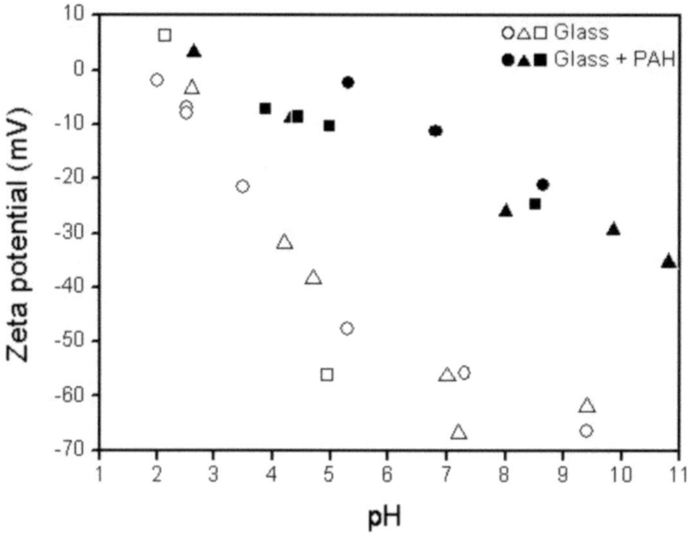

Fig. 3. Zeta potential (mV) of glass (open symbols) and PAH-conditioned glass (dark symbols) as a function of pH in 0.01 M KNO_3. Three experiments were performed on samples prepared independently.

3.2 Nanostructured surfaces with monomodal roughness

Figure 4a shows representative micrographs of samples prepared with substrates conditioned as described above (PAH 10^{-5} M, pH 11, I 10^{-2} M), incubated with 470 nm particles (0.1%, pH 7, I 10^{-3} M), rinsed 6 times with water and dried with a nitrogen flow. The adhering particles were under the form of aggregates, usually bidimensional along the surface plane, sometimes tridimensional (i.e. forming multilayers). Attempts were made to improve the homogeneity of the colloidal particle distribution, keeping the colloidal particle treatment unchanged but using polycations which differ according to functional groups, hydrophobicity and size, and changing polycation treatment conditions : (i) PAH, PLL, PEI, lPDDA and hPDDA; 10^{-7} M at pH 3 and 10^{-5} M at pH 11; I $10^{-3}$and 10^{-1} M; (ii) PAH and lPDDA; 10^{-7} and 10^{-5} M; pH 3, 7 and 11; I 10^{-3}, 10^{-2} and 10^{-1} M. In summary, polycation adsorption at pH 7 or 11 and low ionic strength (10^{-3}, 10^{-2} M) provided a higher degree of coverage by colloidal particles; PAH was as efficient as, or better than other polymers tested; a polycation concentration of 10^{-5} M was more efficient than 10^{-7} M. However in all cases, the adhering particles were under the form of aggregates as described above.

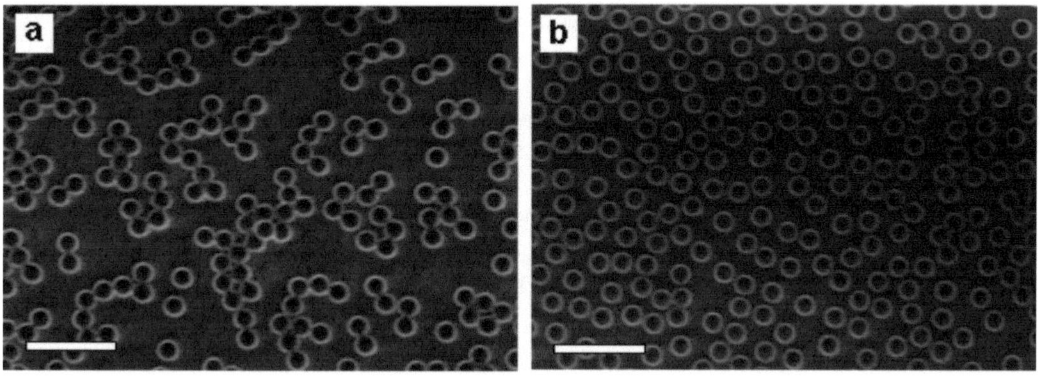

Fig. 4. SEM images obtained on PAH-conditioned (10^{-5}M, pH=11, I=10^{-2}M) glass samples, submitted to latex particle adhesion (470 nm in diameter, 0.1%, pH=7, I=10^{-3}M), rinsed with water and dried under a gentle nitrogen flow (a) or rinsed with water, then with isopropanol and air-dried (b). Scale bars: 2 µm.

A more homogeneous particle distribution was obtained when the sample was rinsed first with water then with isopropanol and left to dry in air overnight, as shown in Figure 4b. Both images of Figure 4 present the same degree of particle coverage (~ 40 %), but in image b, no aggregation is present and the particle distribution is more random. The influence of rinsing and drying procedures on the layer of latex particles may be due to their gathering together by the movement of the liquid film or by the nitrogen flow. In order to avoid particle aggregation at high coverage, Hanarp et al. (2001) attempted to retain particles in place during drying by adsorbing smaller silica particles between the latex particles or by heating the samples with adhering particles in boiling water in order to deform the particles and increase the contact area with the substrate. In the present work, rinsing with isopropanol was found to be efficient (Figure 4b) ; this is attributed to the reduction of capillary forces owing to the 3 times lower surface tension of isopropanol compared to water ((21.7 mN/m compared to 73.0 mN/m (Weast, 1972).
Particle adhesion was further studied using substrates conditioned with PAH as described above, rinsing 3 times with water and 3 times with isopropanol, followed by air drying. Figure 5 presents representative results obtained by incubating the conditioned substrate during 2 hours with the colloids at 0.1 % concentration, using 3 pH values and 2 ionic strengths. At pH 7 with a low ionic strength and at pH 11, substrates were covered by a homogeneous layer of particles. In contrast, a high degree of aggregation was observed at pH 3 whatever the ionic strength, and at pH 7 with a high ionic strength.
Figure 6 (a, b) presents micrographs obtained after adhesion of the two kinds of latex particles separately (470 nm or 65 nm; 0.1%, pH 7, I 10^{-3} M), with isopropanol rinsing and air drying. Two counting methods were used in order to assess the degree of substrate coverage. Binarization and automatic processing of SEM images allow counting to be made more rapidly and on a large number of images in comparison with manual counting.

However, it involves operations which could possibly create systematic variations. Binarization of SEM images gave degrees of coverage of 50.2 ± 5 % and 14.6 ± 4 % on surfaces layered with 470 and 65 nm particles, respectively. A consistent degree of substrate coverage was obtained using manual counting (470 nm : 45.2 ± 2 %; 65 nm : 10.7 ± 1 %). The differences between the two modes of counting were thus in the range of the standard deviation. Samples with adherent particles were also analyzed by XPS (Table 1). For samples 7 and 8 , 470 nm particles were used and the degree of coverage was 53 and 27 %, respectively; for sample 9, 65 nm particles were used and the degree of coverage was 15 %, as assessed by SEM image analysis. As compared with PAH conditioned substrates with no adhering latex, these three samples gave a higher concentration of carbon, a lower concentration of silicon, oxygen and nitrogen and indicated the presence of sulfur, characterized by a S 2p peak (showing the $2p_{3/2}$ and $2p_{1/2}$ components with S $2p_{3/2}$ observed at 169 eV), characteristic of sulfate and due to the latex surface.

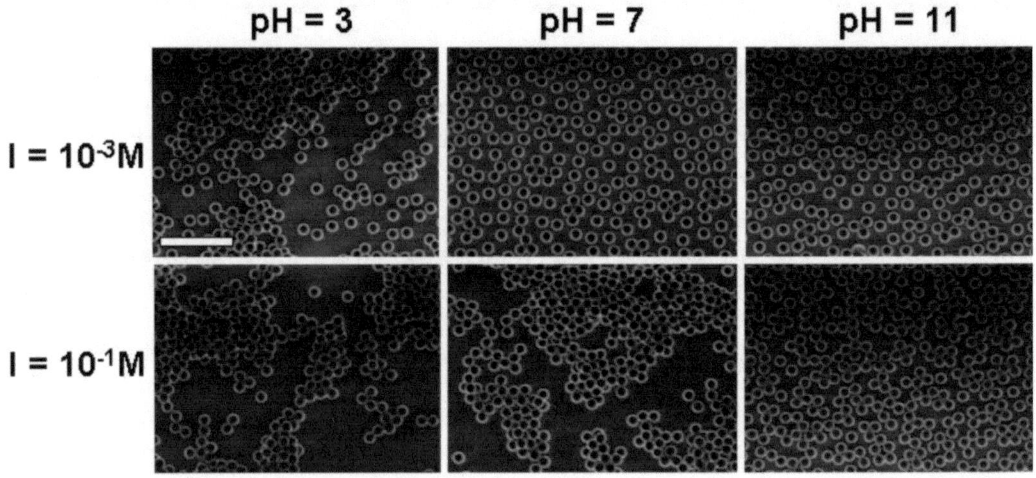

Fig. 5. SEM images obtained on PAH-conditioned (10^{-5}M, pH=11, I=10^{-2}M) glass samples, submitted to latex particle adhesion in different conditions of pH and ionic strength (470 nm in diameter particles, 0.1%), rinsed with water, then with isopropanol and air dried. Scale bars: 2 µm.

A variation of the duration of incubation of the PAH-conditioned substrate with the 470 nm latex (1 to 6 hours, 1 to 6 days) showed that an incubation time of at least 2 hours was required in order to obtain these degrees of coverage. However no increase of coverage was obtained when incubating for longer periods of times. Using a 1 % latex concentration instead of 0.1 % did not provide any rise of the degree of coverage (data not shown).

Obtaining a layer of adhering particles deserves discussion, considering not only interfacial interactions but also the amount of colloidal particles involved and mass transfer. Table 2 presents data computed for the experimental conditions of the treatments with 0.1 % colloidal suspensions. Similar data for PAH 10^{-5} M solution are given for comparison, considering the polycation as a non hydrated sphere of 3.0 nm radius; note that the hydrodynamic radius at pH 7.4 is about 5 times larger (Adamczyk et al., 2006).

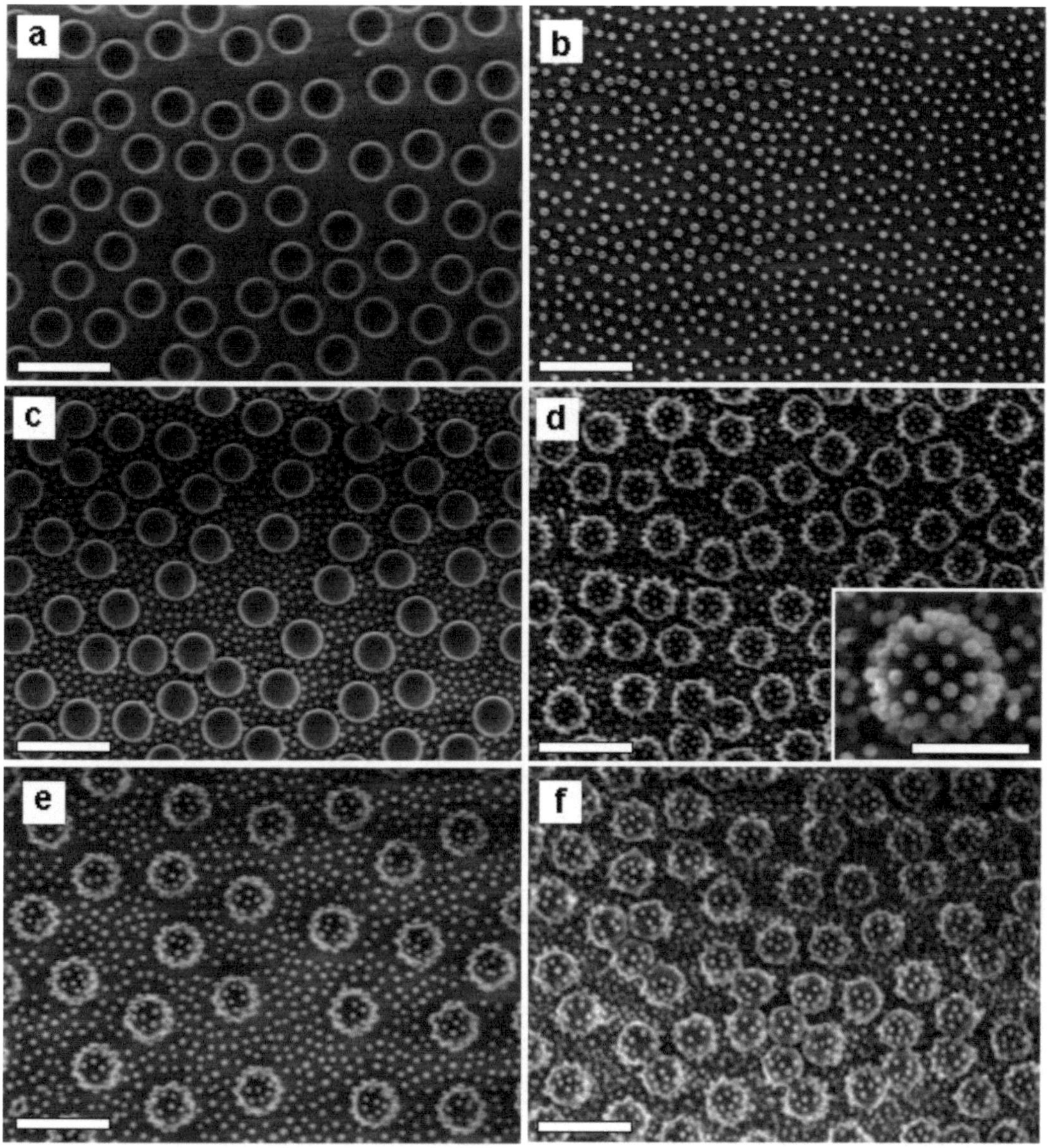

Fig. 6. SEM images obtained on PAH-conditioned (10^{-5}M, pH=11, I=10^{-2}M) glass samples, submitted to adhesion of latex particles (0.1%, pH=7, I=10^{-3}M, rinsing with water then with isopropanol) with 470 nm (a) and 65 nm diameter (b), and submitted to sequential adhesion of 470 nm and 65 nm particles without (c, bimodal I roughness) and with (d, e, f, bimodal II roughness) PAH adsorption after adhesion of the 470 nm particles. Image f was obtained on a sample prepared in the same experimental set as d, then submitted to immersion in PBS at 37°C for 24h. Images d and e were obtained on samples fabricated independently. The inset shows an enlargement of structures present on surfaces with bimodal II roughness. Scale bars: 1 μm. Scale bar inset: 500 nm.

- Area A occupied per particle in hexagonal close packing (3.46 a^2, a being the particle radius).
- Amount Q of particles required to cover the surface with a monolayer in hexagonal close packing.
- Concentration C of particles in the liquid phase, which, if expressed in ml^{-1}, is also the amount of particles involved in the preparation.
- Thickness e of the liquid layer containing the amount of particles required to make a monolayer.
- Diffusion coefficient D of the particles ($D = kT / 6\pi\eta a$), where k, T and η are the Boltzmann constant, the temperature and the viscosity, respectively.
- Time t_d required to produce a monolayer if the rate of adsorption is entirely controlled by diffusion through a concentration gradient created by adsorption itself (Crank, 1957): $t_d = \pi Q^2 / 4 C^2 D$.

It appears that the particles are in large excess of what is needed to make a monolayer. Mass transport by diffusion is fast for polycations and 65 nm latex particles. In contrast, a period of one day is expected to be needed for insuring the formation of a monolayer of 470 nm particles if diffusion is the rate controlling process. Thus sedimentation or convection should contribute to bringing the 470 nm particles to the surface. Considering a density of 1.05 g/cm², the rate of sedimentation v_s of 470 nm particles is about 6.0 nm/s according to Stokes law. Accordingly the time t_s required to produce a monolayer if the rate is entirely controlled by sedimentation is $t_s = e / v_s$ = 13 hours. Actually the amount of adhering 470 nm particles does not increase significantly if the incubation time is extended from 1 hour to 6 days, suggesting that convection due to manipulation insures mass transport. The data of Table 2 also help to figure out in what limits it would be possible to control the adsorption of polycations and small latex particles by playing with incubation time and concentration.

	PAH **10^{-5} M**	**Latex 65** **0.1 %**	**Latex 470** **0.1 %**
A (nm^2 $part^{-1}$)	32	3.7 10^3	1.91 10^5
Q (part. cm^{-2})	3.1 10^{12}	2.7 10^{10}	5.2 10^8
C (part. cm^{-3})	6.0 10^{15}	7.0 10^{12}	1.8 10^{10}
e (µm)	5.2	39	284
D (cm^2s^{-1})	7.2 10^{-7}	6.7 10^{-8}	9.3 10^{-9}
t_d	0.29 s	2.9 min	19 hours

Table 2. Particle characteristics relevant to experimental conditions. See text for details.

While mass transport is not a limiting factor for adhesion of 65 nm particles, the density of the monolayer remains fairly low (10 to 15 % degree of coverage). For this latex at the ionic strength of 10^{-2} M, the product κa, where κ is the inverse of the Debye length, is about 10. For this κa value, maximum coverages of about 35 % were reported for positively charged latex particles on mica (Johnson and Lenhoff, 1996) and for negatively charged latex particles on titanium oxide modified by successive adsorption of PDDA, polystyrene sulfonate and aluminium chloride hydroxide (Hanarp et al., 2001), in agreement with predictions based on repulsion between the particles considered as hard spheres. It is thus reasonable to consider that the coverage by the 65 nm latex is controlled by double layer repulsion between particles. For this latex, a surface coverage of 10 to 15 % corresponds to an apparent area of 37 10^3 to 24 10^3 nm^2/ particle, corresponding to an occupational diameter of 207 to 166 nm. The difference (142 to 101 nm) with the real particle size is larger that the latter, which demonstrates that the degree of coverage is limited by particle-particle repulsion.

If the particles were in hexagonal close packing, the area occupied per particle would be as presented in Table 2. A surface coverage of 45 % for the 470 nm latex thus corresponds to an apparent area per particle of 4.2 10^5 nm^2 and an occupational diameter of 700 nm. The difference of 230 nm between the occupational diameter and particle size is lower than the real particle diameter. Accordingly the degree of coverage is limited by the space available, considering a random distribution of particles. Consequently, the degree of coverage would be limited by the space available for 470 nm particles and by electrostatic repulsion for the 65 nm latex. Hanarp (2003) used the ionic strength (10^{-5} to 10^{-2} M), to control the density of 110 nm polystyrene particles adhering on titanium oxide conditioned with a solution of aluminium chloride hydroxide. Our data indicate that this method would not work for particles above 200 nm diameter.

Figure 5 shows aggregation of 470 nm particles when using a 10^{-1} M ionic strength at pH 3 or 7, or when using a 10^{-3} M ionic strength at pH 3. The surface charge of colloids used here is not affected by pH as confirmed by Schulz et al. (1994a, 1994b), who used latex particles of 131 nm from the same provider and with the same surface specifications as the particles used here. If the aggregates were formed in the suspension, the adhering amount would be expected to increase with time in the range of a few hours or when increasing the concentration from 0.1 to 1 %, in contrast with observations. This suggests that aggregation may take place at the sol - substrate interface owing to low particle - particle repulsion, in addition to the tendency to gather together induced by subsequent solvent evaporation. The aggregation may also be favored by partial desorption of the polycation, as it is enhanced at a pH 3 and 7, at which the polycation is more highly charged (Choi and Rubner, 2005).

3.3 Nanostructured surfaces with bimodal roughness

Surfaces with bimodal I roughness (Figure 6c) were created by adding 65 nm particles subsequently to the 470 nm particles (cf Figure 1c,d). The degree of substrate coverage with 470 nm particles was of 46.9 ± 1 % on this sample. This approach is similar to that of Takeshita et al. (2004) using poly(ethylene terephthalate) conditioned with PAH and adhesion of 350 nm carboxylated polystyrene latex particles followed by 100 nm particles. However, the layer obtained in that case was not regular, despite sonication of the solution to prevent latex aggregation. We attribute this to the fact that the samples were rinsed with water before drying.

Surfaces with bimodal II roughness (Figure 6d) were prepared by adsorbing PAH after the formation of the first adhering layer (cf Figure 1b,c,e,f). In this way, 65 nm particles were adhering not only to the glass substrate but also to the 470 nm particles, providing raspberry-like structures. This procedure was repeated five times independently and 3 to 15 samples were prepared in each experiment, demonstrating the repeatability as well as the reproducibility of the method. Figure 6 (d, e) presents the extreme results obtained, corresponding to degrees of coverage by 470 nm particles of 50 ± 3 % and 22 ± 1 %, respectively.

Owing to the possible interest of nanostructured surfaces in the field of biointerfaces, samples with adhering particles were incubated in phosphate buffer saline (the main constituent of culture media) for 24 h at 37°C, followed by rinsing with water, rinsing with isopropanol and drying. The result obtained with a bimodal II roughness is presented in Figure 6f. The observed morphology is similar to that of the sample not exposed to buffer (Figure 6d), demonstrating the robustness of the protocol and of the nanostructured surface obtained.

Recently published works aimed at creating this kind of roughness using different approaches. Suspensions of raspberry-like particles were prepared by styrene polymerization on silica particles (Perro et al., 2006; Reculusa et al., 2002). In another study, silica particles having different sizes and bearing functional groups were firstly synthesized independently and then mixed to react together. The obtained aggregates were then grafted on a specific substrate to obtain a dual-size roughness surface (Xiu et al., 2006). In another approach, a layer of silica particles in hexagonal close packing was created on a substrate, and gold nanoparticles were formed on the top of the silica spheres by sputtering (Ming et al., 2005). The method used in the present work has several advantages, such as the use of components which are commercially available (polycation, latex particles), and a simple procedure (sequential steps of polycation adsorption and colloid adhesion) which does not require sophisticated devices or complex reactions. The protocol could be extended to other particle sizes for obtaining a broader panel of roughness, the density of the two types of particles could be tuned, and architectures could be elaborated with more than two particle sizes. The bimodal surfaces of type II roughness mimic the particular topography observed on Lotus leaf. A superhydrophobic surface is thus expected to be obtained after treatment with compounds conferring a low surface energy (Bravo et al., 2007).

4. Conclusion

Nanostructured surfaces were fabricated through assembly of PAH and polystyrene latex particles. Thereby, different types of roughness were created, with a single layer of particles (65 or 470 nm), a layer of two types of particles (65 and 470 nm) - bimodal roughness of type I - or a layer of raspberry-like relief features (65 nm on 470 nm) – bimodal roughness of type II. The best conditions for glass conditioning by PAH were a high pH (11) and a low ionic strength (10^{-2}M). A neutral or alkaline pH with a low ionic strength was satisfactory for the adhesion of the colloidal particles on PAH-conditioned glass. Terminating the rinsing procedure with isopropanol before air drying was needed to avoid particle aggregation due to capillary forces. Incubation in PBS, a buffer solution mimicking the electrolyte composition of biological fluids, did not alter the structures obtained.

After PAH conditioning, the thickness of the adsorbed layer detected on glass was in the range of 2.5 to 3.5 nm ; this layer contained an appreciable amount of adventitious organic contaminants. The degree of ionization of PAH at the outermost part of the adsorbed layer was low and did not provoke a surface charge reversal. This indicates that the colloid particles were not attracted by long distance forces but rather suggests a change of PAH protonation with a redistribution of counterions when colloidal particles approached the PAH-modified glass.

The control of the degree of coverage by adhering particles for performing colloidal lithography may be pursued by playing with interfacial interactions, and thus with pH and ionic strength. On the other hand it may be pursued by playing with mass transfer:

substrate orientation (upward, downward, vertical), convection, particle concentration and contact time. The selection of the best approach depends primarily on the particle size, which is critical in the range of 100 nm, and secondarily on particle density and on the desired degree of coverage.

5. Acknowledgements

The support of the Foundation for Training in Industrial and Agricultural Research (FRIA), of the Belgian National Foundation for Scientific Research (FNRS), of the Région Wallonne and of the Federal Office for Scientific, Technical and Cultural Affairs (Interuniversity Poles of Attraction Program) is gratefully acknowledged.

6. References

Adamczyk, Z., Zembala, M., & Michna, A. (2006). Polyelectrolyte adsorption layers studied by streaming potential and particle deposition. *J. Colloid Interface Sci.*, 303, 353-364

Adamczyk, Z., Zembala, M., Kolasinska, M., & Warszynski, P. (2007) Characterization of polyelectrolyte multilayers on mica and oxidized titanium by steaming potential and wetting angle measurements. *Colloid Surf. A: Physicochem. Eng. Asp.*, 302, 455-460

Agheli, H., Malmstrom, J., Hanarp, P., & Sutherland, D.S. (2006). Nanostructured biointerfaces. *Mater. Sci. Eng. C: Biomimetic Supramol. Syst.*, 26, 911-917

Barthlott, W., & Neinhuis C. (1997). Purity of the sacred lotus, or escape from contamination in biological surfaces. *Planta,* 202, 1-8

Bertrand, P., Jonas, A., Laschewsky, A., & Legras R. (2000). Ultrathin polymer coatings by complexation of polyelectrolytes at interfaces: Suitable materials, structure and properties. *Macromol. Rapid Commun.*, 21, 319-348.

Boonaert, C.J.P., Dupont-Gillain, C.C., Dengis, P.B., Dufrêne Y.F., & Rouxhet, P.G. (1999). Cell separation, flocculation, In: *Encyclopedia of bioprocess technology: Fermentation, biocatalysis, and bioseparation,* M.C. Flickinger, S.W. Drew (Eds.), pp. 531-548, John Wiley & Sons, Inc., New York

Boonaert, C.J.P., Dufrêne, Y.F., & Rouxhet, P.G. (2002). Adhesion (primary) of microorganisms onto surfaces, In: *Encyclopedia of environmental microbiology,* G. Bitton (Ed.), pp. 113-132, John Wiley & Sons Inc., New York

Bravo, J., Zhai, L., Wu, Z., Cohen, R.E., & Rubner M.F. (2007). Transparent superhydrophobic films based on silica nanoparticles. *Langmuir,* 23, 7293-7298

Caillou, S. , Gerin, P.A., Nonckreman, C.J., Fleith, S., Dupont-Gillain, C.C., Landoulsi, J., Pancera, S.M., Genet , M.J., & Rouxhet, P.G. (2008). Enzymes at solid surfaces: nature of the interfaces and physico-chemical processes. *Electrochim. Acta,* 54, 116-122

Changui, C., Doren, A., Stone, W.E.E., Mozes, N, & Rouxhet, P.G. (1987). Surface properties of polycarbonate and promotion of yeast cell adhesion. *Journal de Chimie Physique et de Physico-Chimie Biologique*, 84, 275-281

Chen, K.M., Jiang, X., Kimerling, L.C., & Hammond, P.T. (2000). Selective self-organization of colloids on patterned polyelectrolyte templates. *Langmuir*, 16, 7825-7834

Choi, J., & Rubner, M.F. (2005). Influence of the degree of ionization on weak polyelectrolyte multilayer assembly. *Macromolecules*, 38, 116-124

Claesson, P.M., Poptoshev, E., Blomberg, E., & Dedinaite, A. (2005). Polyelectrolyte-mediated surface interactions. *Adv. Colloid Interface Sci.*, 114-115, 173-187

Crank, J. (1957). *The mathematics of diffusion*, Oxford University Press, p. 30-31

Davies, R.J., Dix, L.R., & Toprakcioglu C. (1989). Adsorption of poly-L-lysine to mica powder. *J. Colloid Interface Sci.*, 129, 145-152

Decher, G., & Hong, J.D. (1991). Build up of ultrathin multilayer films by a self-assembly process: Consecutive adsorption of anionic and cationic bipolar amphiphiles on charged surfaces. *Macromol. Chem. Macromol. Symp.*, 46, 321-327

Dekeyser, C.M., Biltresse, S., Marchand-Brynaert, J., Rouxhet, P.G., & Dupont-Gillain, C.C. (2004). Submicrometer-scale heterogeneous surfaces by PS-PMMA demixing. *Polymer*, 45, 2211-2219

Denis, F.A., Hanarp, P., Sutherland, D.S., & Dufrene, Y.F. (2002). Fabrication of nanostructured polymer surfaces using colloidal lithography and spin-coating. *Nano Lett.*, 2, 1419-1425

Elgersma, A.V., Zsom, R.L.J., Lyklema, J., & Norde, W. (1992). Kinetics of single and competitive protein adsorption studied by reflectometry and streaming potential measurements. *Colloid Surf.*, 65, 17-28

Fustin, C.A., Glasser, G., Spiess, H.W., & Jonas, U. (2004). Parameters influencing the templated growth of colloidal crystals on chemically patterned surfaces. *Langmuir*, 20, 9114-9123

Gerin, P.A., Dengis, P.B., & Rouxhet, P.G. (1995). Performance of XPS analysis of model biochemical coumpounds. *Journal de Chimie Physique*, 92, 1043-1065

Hanarp, P., Sutherland, D.S., Gold, J., & Kasemo B. (2001). Influence of polydispersity on adsorption of nanoparticles. *J. Colloid Interface Sci.*, 241, 26-31

Hanarp, P., Sutherland, D.S., Gold, J., & Kasemo, B. (2003). Control of nanoparticle film structure for colloidal lithography. *Colloid Surf. A: Physicochem. Eng. Asp.*, 214, 23-36.

Johnson, C.A., & Lenhoff, A.M. (1996). Adsorption of Charged Latex Particles on Mica Studied by Atomic Force Microscopy. *J. Colloid Interface Sci.*, 179, 587-599

Lafuma, F. (1996). Mechanisms of flocculation and stabilisation of suspensions by organic polymers, In: *Paper Chemistry*, J.C. Roberts (Ed.), Chapman & Hall, London

Li, J., Luan, S., Huang, W., & Han, Y. (2007). Colloidal crystal heterostructures by a two-step vertical deposition method. *Colloid Surf. A: Physicochem. Eng. Asp.* 295, 107-112

Li, Z., Ravaine, V., Ravaine, S., Garrigue, P., & Kuhn, A. (2007). Rapsberry-like gold microspheres: preparation and electrochemical characterization. *Adv. Functional Materials*, 17, 618-622

Lindquist, G.M., & Stratton, R.A. (1976). The role of polyelectrolyte charge density and molecular weight on the adsorption and flocculation of colloidal silica with polyethylenimine. *J. Colloid Interface Sci.*, 55, 45-59

Meszaros, R., Varga, I., & Gilanyi, T. (2004). Adsorption of poly(ethyleneimine) on silica surfaces: Effect of pH on the reversibility of adsorption. *Langmuir*, 20, 5026-5029

Ming, W., Wu, D., van Benthem, R., & de With, G. (2005). Superhydrophobic films from raspberry-like particles. *Nano Lett.*, 5, 2298-2301

Nonckreman, C.J., Fleith, S., Rouxhet, P.G., & Dupont-Gillain, C.C. (2010). Competitive adsorption of fibrinogen and albumin and blood palatelet adhesion on surfaces modified with nanoparticles and/or PEO. *Colloids and Surfaces B: Biointerfaces, 77,* 139-149

Patankar, N.A. (2004). Mimicking the Lotus Effect: Influence of double roughness structures and slender pillars. *Langmuir,* 20, 8209-8213

Perro, A., Reculusa, S., Bourgeat-Lami, E., Duguet, E. , & Ravaine, S. (2006). Synthesis of hybrid colloidal particles: From snowman-like to raspberry-like morphologies. *Colloid Surf. A: Physicochem. Eng. Asp.,* 284-285, 78-83

Petrov, A.I., Antipov, A.A., & Sukhorukov, G.B. (2003). Base-acid equilibria in polyelectrolyte systems: From weak polyelectrolytes to interpolyelectrolytes complexes ans multilayered polyelectrolyte shells. *Macromolecules,* 36, 10079-10086.

Reculusa, S., Poncet-Legrand, C., Ravaine, S., Mingotaud, C., Duguet, E., & Bourgeat-Lami, E. (2002). Syntheses of raspberrylike silica/polystyrene materials. *Chem. Mater.,* 14, 2354-2359

Roberts, J.C. (1996). The surface chemistry of paper and the paper-making system, In: *The Chemistry of Paper,* J.C. Roberts (Ed.), The Royal Society of Chemistry, Letchworth.

Rouxhet, P.G., Doren, A., Dewez, J.L., & Heuschling, O. (1993). Chemical composition and physico-chemical properties of polymer surfaces. *Prog. Org. Coat.,* 22, 327-344

Schaak, R.E., Cable, R.E., Leonard, B.M., & Norris, B.C. (2004). Colloidal crystal microarrays and two-dimensional superstructures: a versatile approach for patterned surface assembly. *Langmuir,* 20, 7293-7297

Schulz, S.F., & Sticher, H. (1994a). Surface charge densities and electrophoretic mobilities of aqueous colloidal suspensions of latex spheres with different ionizable groups. *Prog. Colloid Polym. Sci.,* 97, 85-88

Schulz, S.F., Gisler, T., Borkovec, M., & Sticher, H. (1994b). Surface charge on functionalized latex spheres in aqueous colloidal suspensions. *J. Colloid Interface Sci.,* 164, 88-98

Tagliazucchi, M., Calvo, E.J., & Szleifer, I. (2008). Redox and acid-base coupling in ultrathin polyelectrolyte films. *Langmuir,* 24, 2869-2877

Takeshita, N., Paradis, L.A., Oener, D., McCarthy, T.J., & Chen, W. (2004). Simultaneous tailoring of surface topography and chemical structure for controlled wettability. *Langmuir,* 20, 8131-8136

Tanuma, S., Powell, C.J., & Penn, D.R. (1997). Calculations of electron inelastic mean free paths (IMFPs). VI. Analysis of the Gries inelastic scattering model and predictive IMFP equation. *Surf. Interface Anal.,* 25, 25-35

Van Haecht, J.L., Bolipombo, M., & Rouxhet, P.G. (1985). Immobilization of Saccharomyces cerevisiae by adhesion: treatment of the cells by aluminum ions. *Biotechnol. Bioeng.,* 27, 217-224

Weast, R.C. (1972). *Handbook of Chemistry and Physics,* 52nd ed., The Chemical Rubber C.O., Cleveland.

Wenzel, R.N. (1936). Resistance of solid surfaces to wetting by water. *J. Ind. Eng. Chem.*, 28, 988-994

Wood, M.A. (2007). Colloidal lithography and current fabrication techniques producing in-plane nanotopography for biological applications. *J. R. Soc. Interface*, 4, 1-17

Xiu, Y., Zhu, L., Hess, D.W., & Wong, C.P. (2006). Biomimetic creation of hierarchical surface structures by combining colloidal self-assembly and Au sputter deposition. *Langmuir*, 22, 9676-9681

Yang S., Cai, W., Yang , J., & Zeng, H. (2009). General and simple route to micro/nanostructured hollow-sphere arrays based on electrophoresis of colloids induced by laser ablation in liquid. *Langmuir*, 25, 8287-9291

5

Physical Deposition Assisted Colloidal Lithography: A Technique to Ordered Micro/Nanostructured Arrays

Yue Li[1,3], Shuyan Gao[2], Guotao Duan[1], Naoto Koshizaki[3] and Weiping Cai[1]
[1]Key Laboratory of Materials Physics, Institute of Solid State Physics, Chinese Academy of Sciences, Anhui,
[2]College of Chemistry and Environmental Science, Henan Normal University, Xinxiang
[3]Nanosystem Research Institute (NRI) National Institute of Advanced Industrial Science and Technology (AIST)
[1,2]China
[3]Japan

1. Introduction

Ordered micro/nanostructured arrays have attracted much interest due to their important applications in microfluidic devices, optoelectronic devices, nanophotonics, field emitters, nanogenerators, sensors, nano-biotechnology, surface science, photocatalytic properties etc.[1-11] The traditional routes to create periodic micro/nanostructured arrays are generally divided into two step. Microsized structure arrays are first fabricated by traditional lithographic techniques (e.g. photo-lithography, electron-beam lithography, ion beam lithography, x-ray lithography)[12-15.] as well as soft lithography (e.g. the techniques of replica molding, microcontact printing, micromolding in capillaries)[16-19], the nanostructures are then modified on the microsized units in array,[20] thus hierarchical micro/nanostructured arrays are finally achieved. However, they cannot be afforded due to the high costs and time-consuming in the most laboratories. Recently, the monolayered colloidal crystals (or called colloidal monolayers), ordered monolayer colloidal sphere arrays with hexagonal close-packed lattice structures on a certain substrate by self-assembly,[21-35] can be used to prepare ordered structure arrays. [36-41] It has proved that it is a flexible approach to fabricate the periodic micro or nanostructure arrays (e.g. nanoparticle arrays,[42-49.] nanopore arrays,[50-59] hollow sphere arrays[60-65]) based on colloidal monolayer templates by the different routes, solution/sol-dipping route, electrochemical deposition etc. Their properties are morphology and arrangement parameter dependent. Besides these periodic structure arrays, the colloidal monolayer template also can be used to prepare hierarchical micro/nanostructured arrays. For example, the hierarchical micro/nanostructured polystyrene (PS) sphere/CNTs composite arrays were obtained by wet chemical self assembling;[66-68] hierarchical microsized PS sphere/silver nanoparticle composite arrays or microsized pore/silver nanoparticle arrays were made by thermal deposition of silver precursor;[69,70] gold hierarchical micro/nanostructured particle arrays were created by electrochemical deposition based two step replication of colloidal monolayer template.[71] However, these routes have been developed by basically chemical reaction. They have some disadvantages of impurities on surface of arrays due to incompletely decomposition of precursors, residua of surfactants in self-assembling or electrochemical deposition. Additionally, it is quite difficult to achieve very uniform morphology of hierarchical micro/nanostructure arrays on a large-scale. Another route of colloidal monolayer template combining with physical deposition is expected to resolve these problems. In this chapter, we focus on introducing the recent work to create micro/nanostructured arrays based on colloidal templates with physical deposition (pulsed laser deposition (PLD) and sputtering). The parameters of microstructure or nanostructure can be tuned by periodicities of colloidal templates or experimental conditions of physical deposition. The applications of nanorod arrays with controllable morphology and arrangement parameters in self-cleaning surfaces, enhanced catalytic properties, field emitters etc. are also presented in following section.

2. Pulsed laser deposition assisted colloidal lithography[73,74]

2.1 Method

A polystyrene (PS) colloidal monolayer was first fabricated on a substrate. The desired material was then deposited on this colloidal monolayer substrate by PLD at room temperature and oxygen was introduced into PLD chamber as the background gas. This periodic array has a special hierarchical micro/nanostructure array with a hexagonal-close-packed (hcp) arrangement, which originate from the pattern of colloidal monolayer. In this micro/nanostructure unit in array, the nanorod stands vertically on the microsized PS sphere tops, and nanobranches in each nanorod grow in a radiationlike manner, perpendicular to the PS sphere surface. The detailed experiments are described as following.

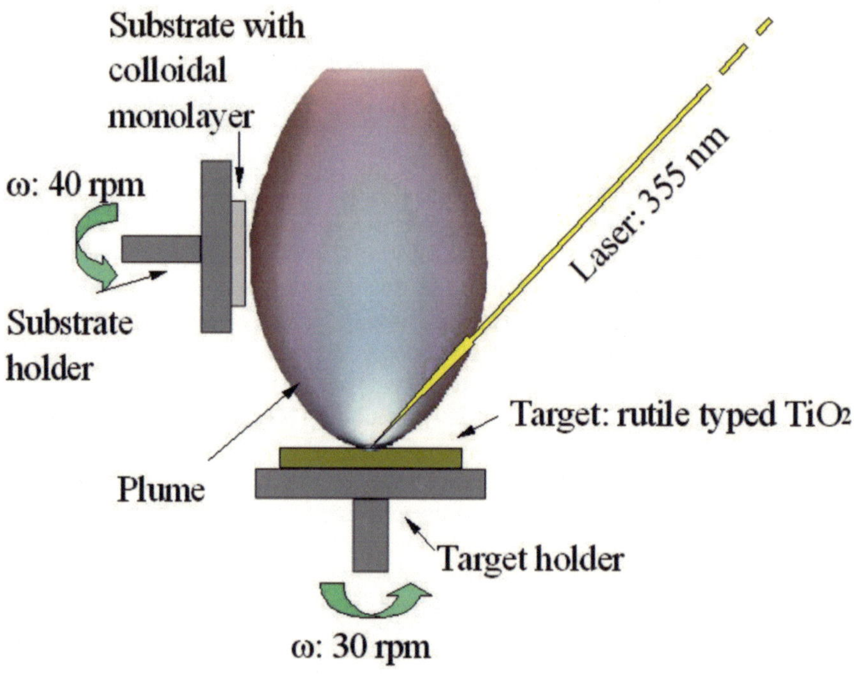

Fig. 1. Schematic illustration of PLD process.

Commercial monodispersed PS spheres dispersed in water with a certain size were purchased from companies. The PS colloidal monolayers were first fabricated on cleaned Si substrates by self-assembly using spin coating. The colloidal monolayer with its supporting substrate was placed in a deposition chamber of PLD, close to the target and at an off-axial position with respect to the target, as shown in Figure 1.A laser beam with a 355 nm wavelength from a Q-switched Nd:YAG laser (Continuum, Precision 8000), operated at 10 Hz with 100 mJ/pulse and a pulse width of 7 ns was applied and focused on the target surface with a diameter of about 2 mm. The desired target, for example, rutile typed titanium dioxide was used for deposition. The substrate and target were rotated at 40 and 30 rpm, respectively. PLD was carried out at a base pressure of 2.66 × 10^{-4} Pa and a background O_2 pressure of 6.7 Pa.

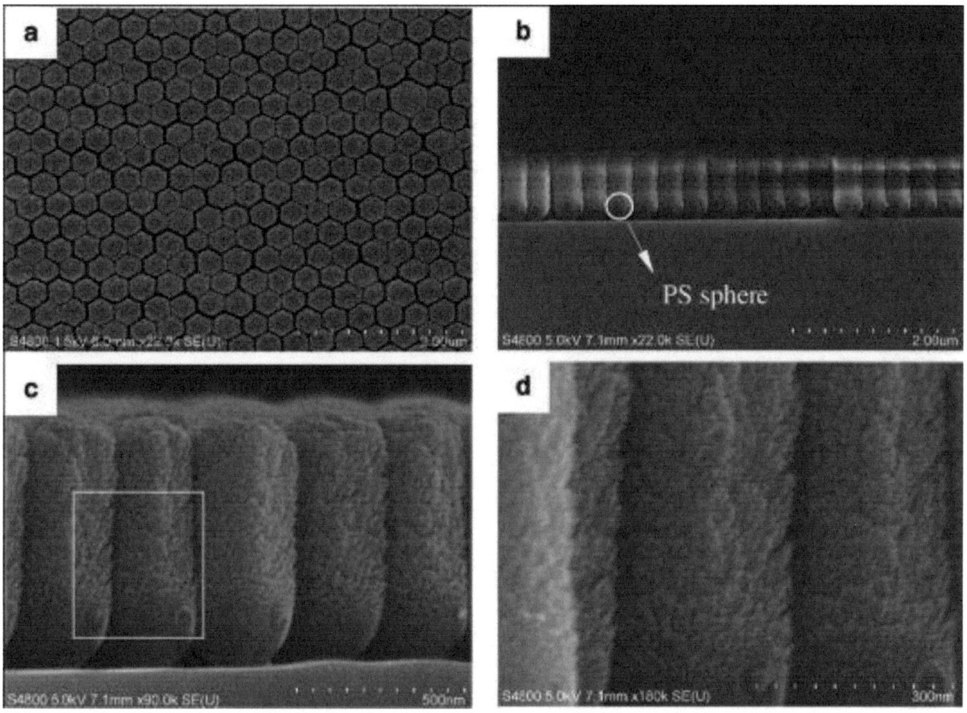

Fig. 2. Morphology of a sample obtained by PLD using a Si substrate with a PS colloidal monolayer coating (PS sphere size: 350 nm; deposition time: 70 min). (a) FESEM image from top view and (b) FE-SEM image of cross-section. (c) and (d) are high-resolution images observed from the side. (d) much higher magnification image of (c).

After deposition, the sample demonstrated a periodic hierarchical micro/nanorod array with an hcp arrangement, as reflected from Figure 2a. Each nanorod consists of two parts: a PS sphere at the bottom and a vertical nanorod on the top of the PS sphere (Figure 2b). The diameter of the nanocolumn was almost the same as that of the PS sphere, 350 nm, and its height was about 870 nm. The nanorod had a very rough structure on the surface and was composed of many nanobranches, according to the high-resolution images of the side view (Figure 2 c, d). TEM observation from the top of the nanorod arrays reflects that each nanorod consists of radiation-shaped nanobranches emanating from the center (Figure 3a).

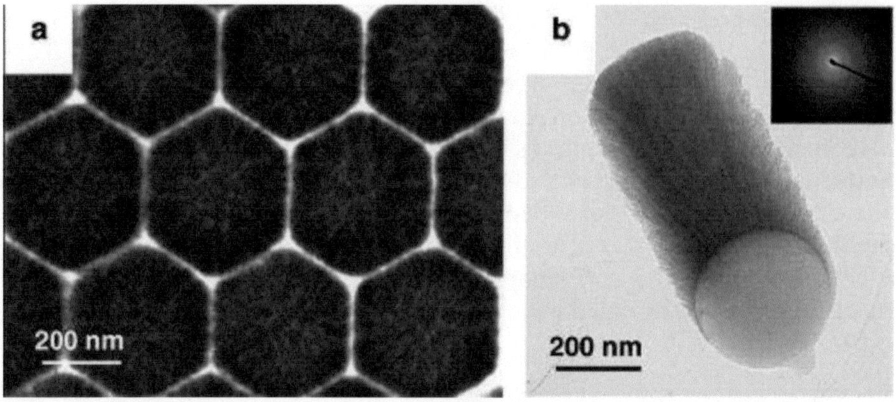

Fig. 3. Corresponding TEM images of the sample in Figure 2. (a) Periodic nanorod array observed from the top. (b) Single nanorod observed from the side. The inset in (b) is the corresponding electron diffraction pattern.

Fig. 4. a, b: FE-SEM images of a CuO hierarchical micro/nanostructured array obtained by combining the PS colloidal monolayer and PLD process. (PS sphere size 350 nm, deposition time 2 h, ambient oxygen pressure during deposition 6.7 Pa). (a) top-view image; (b) section view. Scale bars in parts a and b indicate 500 nm. c, d: TEM images of a CuO hierarchical micro/nanostructured array: (c) TEM image from the top; (d) TEM image of several separated units from the periodic array and the corresponding selected area electron diffraction (SAED) pattern.

The TEM image of a single nanorod also clearly displays that the nanorod consists of a PS sphere at bottom and a nanocolumn on the sphere surface. The nanorod possesses nanobranch-structures, which grow almost vertically on the PS sphere surface (Figure 3b). The nanobranch-structures indicate that the nanorod has a hierarchical, porous structure and hence has a high surface area. The selected area electron diffraction (SAED) pattern shows that the deposited materials on PS sphere surfaces by PLD are amorphous. Besides TiO_2 amorphous hcp nanocolumn arrays, the presented strategy can be extended to the fabrication of similar amorphous structures of SnO_2, WO_3, C, and so forth, just by changing the corresponding target in the PLD process.

Additionally, some materials, e.g. CuO, Fe_2O_3, ZnO are easily crystalline by PLD at room temperature. If the colloidal monolayer is applied as a template, the crystalline CuO, Fe_2O_3, ZnO etc. hierarchical micro/nanostructured arrays can be also obtained. Figure 4 shows the SEM and TEM image CuO crystalline hierarchical micro/nanostructured arrays using colloidal monolayers as templates by PLD. Each arrayed unit is composed of PS sphere at bottom and deposited materials at top. Deposited materials are well crystalline, they do not exhibit round shapes but radially aligned nanocolumns having tips with trigonal pyramidal shapes on the PS sphere.

The deposited CuO nanostructures can be tuned by varying ambient gas pressures during the PLD process. Figure 5 shows the FE-SEM and TEM images of samples achieved by PLD under higher ambient gas pressures during the PLD process using the colloidal monolayers as substrates. When oxygen pressure increased from 6.7 to 26.7 Pa, the morphology did not appreciably change and exhibited similar hierarchical structures as before (Figure 5a). However, when the oxygen pressure increased to 53.3 Pa, the morphology completely changed and was very different from those at lower pressures. The nanocolumn tips on the PS sphere demonstrated imperfect trigonal pyramid shapes, and the tip sizes became much smaller (Figure 5b). According to the corresponding TEM image and SAED pattern (Figure 5c), we find that hierarchical micro-/nanostructures were still observed at such high oxygen pressure, but the crystallization of deposited aligned nanocolumns on the PS sphere becomes worse than that obtained at lower oxygen pressure. When the gas pressure increased to as high as 79.8 Pa, similar hierarchical micro-/nanostructured array was not obtained, and many aggregates of small particles were produced on the colloidal monolayer template (Figure 5d). The XRD spectra of the samples obtained under different oxygen pressures are shown in Figure 6. Strong preferential orientation growth along (002) was observed at the gas pressure of 26.7 Pa. Increasing oxygen pressure led to weakening of this preferential orientation and broadening of X-ray diffraction peaks. This result reflects that deposited materials gradually changed to small nanoparticles from aligned nanocolumn arrays and the particles became much smaller with increasing oxygen pressure during PLD, agreeing with FE-SEM images. When the oxygen pressure increased to very high value, 79.8 Pa, the deposited material completely consisted of small nanoparticles or the aggregates of small nanoparticles, and there was no preferential orientation growth. Because when the gas pressure increases to a high value, the plume is compressed into a smaller space in PLD process, and the possibility of collision among ions or atoms in plasma is greatly enhanced, further resulting in a kinetic energy decrease of ions or atoms, which leads to less crystallization and smaller nanoparticle formation[75].

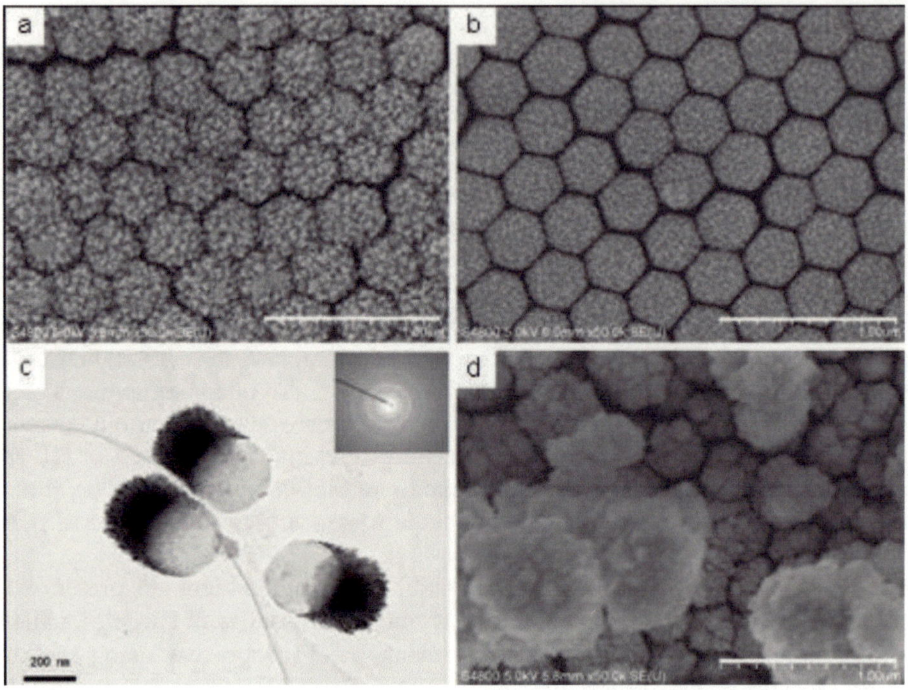

Fig. 5. Images obtained by the different ambient oxygen pressures: (a, b, d) FE-SEM images of the samples obtained under ambient oxygen pressure of 26.7, 53.3, and 79.8 Pa, respectively; (c) TEM image of the sample obtained at 53.3 Pa and the corresponding SAED pattern of several units. Scale bars in (a), (b), and (d): 1 μm.

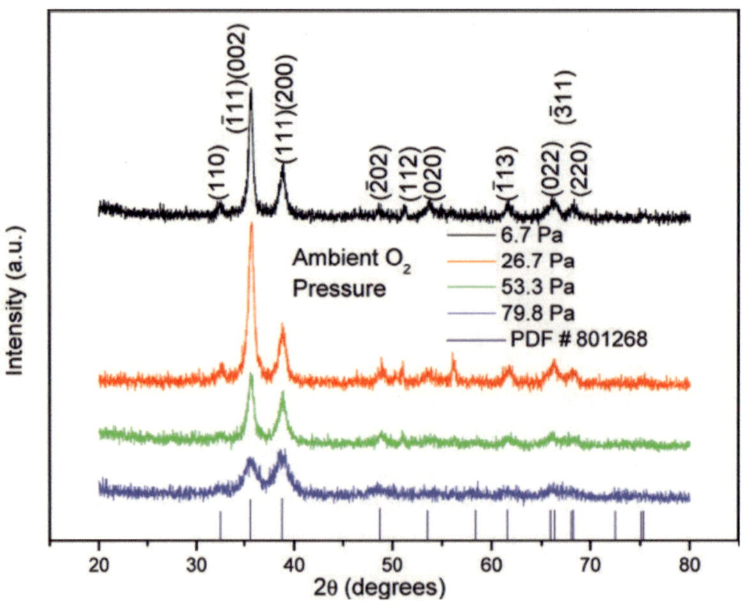

Fig. 6. XRD patterns of the samples obtained under different oxygen pressures.

Similar crystalline hierarchical micro-/nanostructured arrays of Fe_2O_3 and ZnO can be also created by the same route, as shown in Figure 7. Fe_2O_3 nanobelts or ZnO nanocolumns were well aligned on the PS sphere tops, like those of CuO. However, the Fe_2O_3 nanobelt or ZnO nanocolumn tops were not like those of CuO. The slight differences among CuO, Fe_2O_3, and ZnO fine nanostructures are determined mainly by their various chemical and physical properties: a crystal facet of the interface with different energies etc.

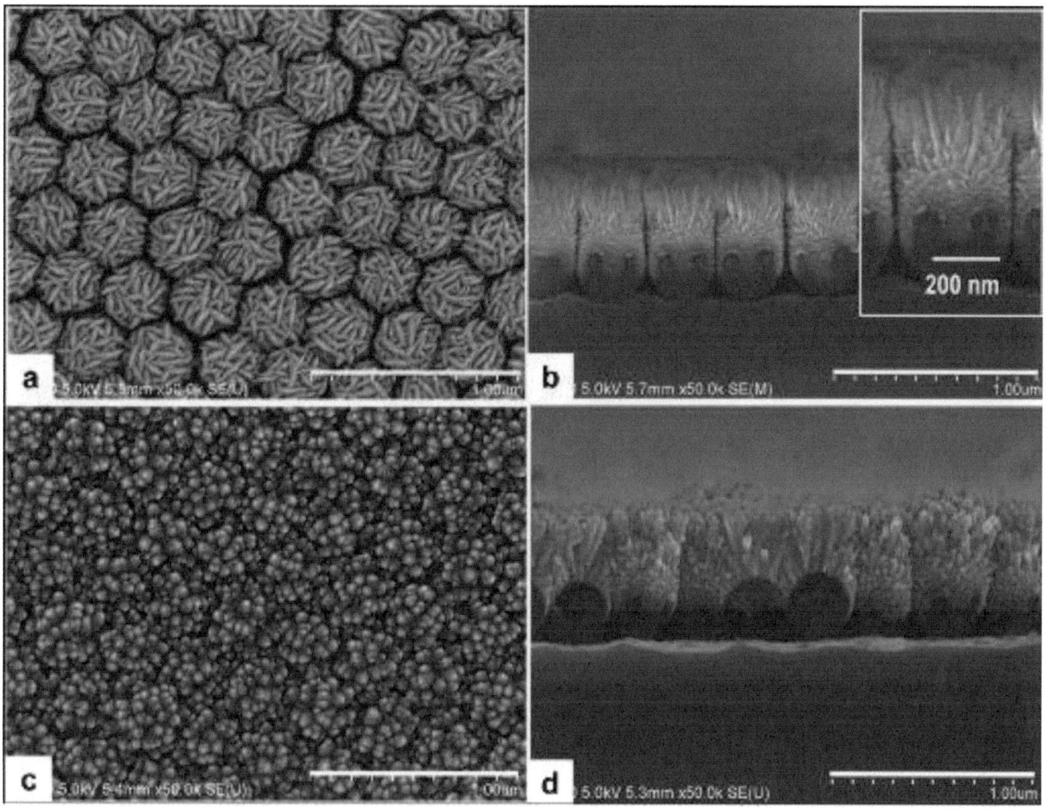

Fig. 7. FE-SEM images of hierarchical micro-/nanostructured arrays of Fe_2O_3 and ZnO. (a, b) Fe_2O_3, oxygen pressure 6.7 Pa, deposition time 1.5 h; (c, d) ZnO. a and c: top views; b and d: side views (oxygen pressure 6.7 Pa and deposition 40 min). The inset in (b): the high magnification image of a single Fe_2O_3 hierarchical micro-/ nanostructure.

In this strategy, the height of micro/nanostructured unit can be obviously controlled by varying deposition time during PLD process, the height will increase with increase of PLD time. From the Figure 8, it can be found that the unit height increases by increasing deposition time from 30 min to 60 min. However, if the deposition time is too long, to say, 180 min, the tops of micor/nanostructured units will aggregate with each other due to strong Van de Waals attraction among units in the deposition process, as shown in Figure 8 (c), (c') and (c''). Additionally, the top of micro/nanostructured unit gradually flattens from convex shape with increasing deposition time, resulting in a weakening shadow effect. Therefore, a continuous film might be formed at top of hierarchical micro/nanostructured array if further increasing deposition time after 180 min.

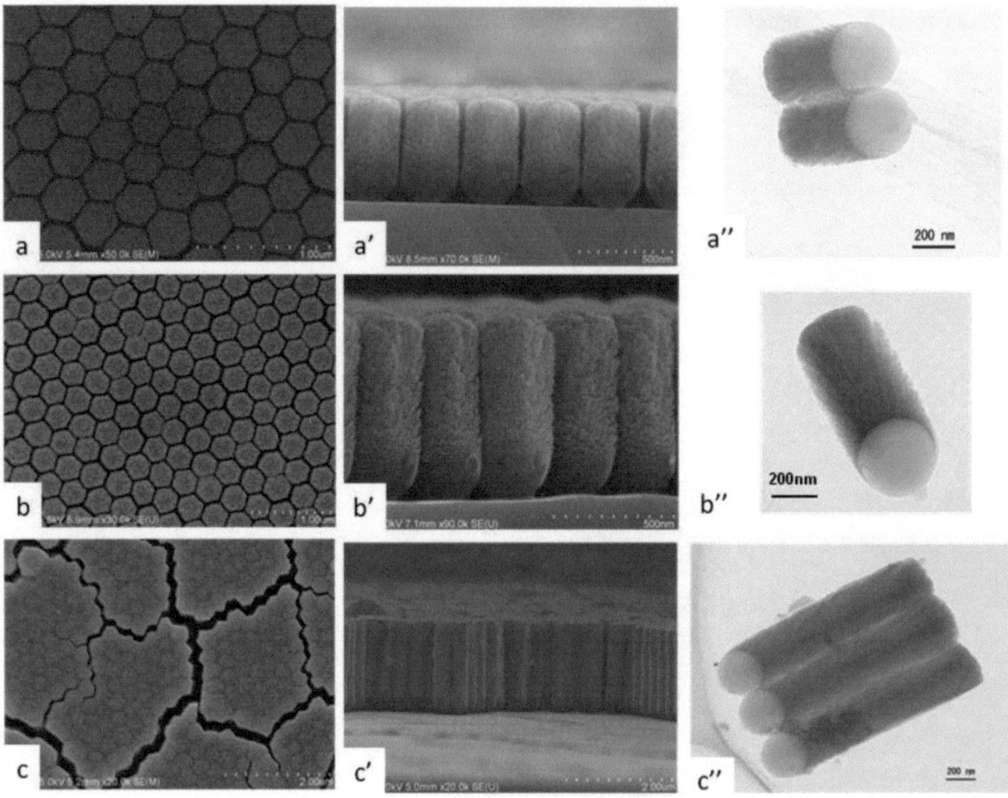

Fig. 8. The height changes of micro/nanostructured unit with increase deposition time. Deposition time: 30 min in (a), (a') and (a''); 60 min in (b), (b') and (b''); 180 min in (c), (c') and (c''). (a), (b) and (c) are SEM images of top view; (a'), (b') and (c') are SEM images of cross-section; (a''), (b'') and (c'') are TEM images of micro/nanostructure units.

The as-prepared hierarchical micro/nanostructured units in periodic arrays are composed of a PS sphere at the bottom and a micro/nano- particle or rod on the top of the PS sphere. If the PS colloidal template is dissolved by an organic solution (CH_2Cl_2), this periodic array could retain its integrity while being peeled from the substrate due to the van der Waals force between the neighboring micro/nanostructured units suspended in the solution. It could then be transferred to any desired substrate (e.g., TEM copper grid) by picking it up using another substrate, as illustrated in Figure 9 and Figure 10. The transferability avoids restrictions on substrates in the fabrication process of hierarchical micro/nanostructured arrays, which is helpful in the design and fabrication of new micro-/nano- devices on any desired substrates.

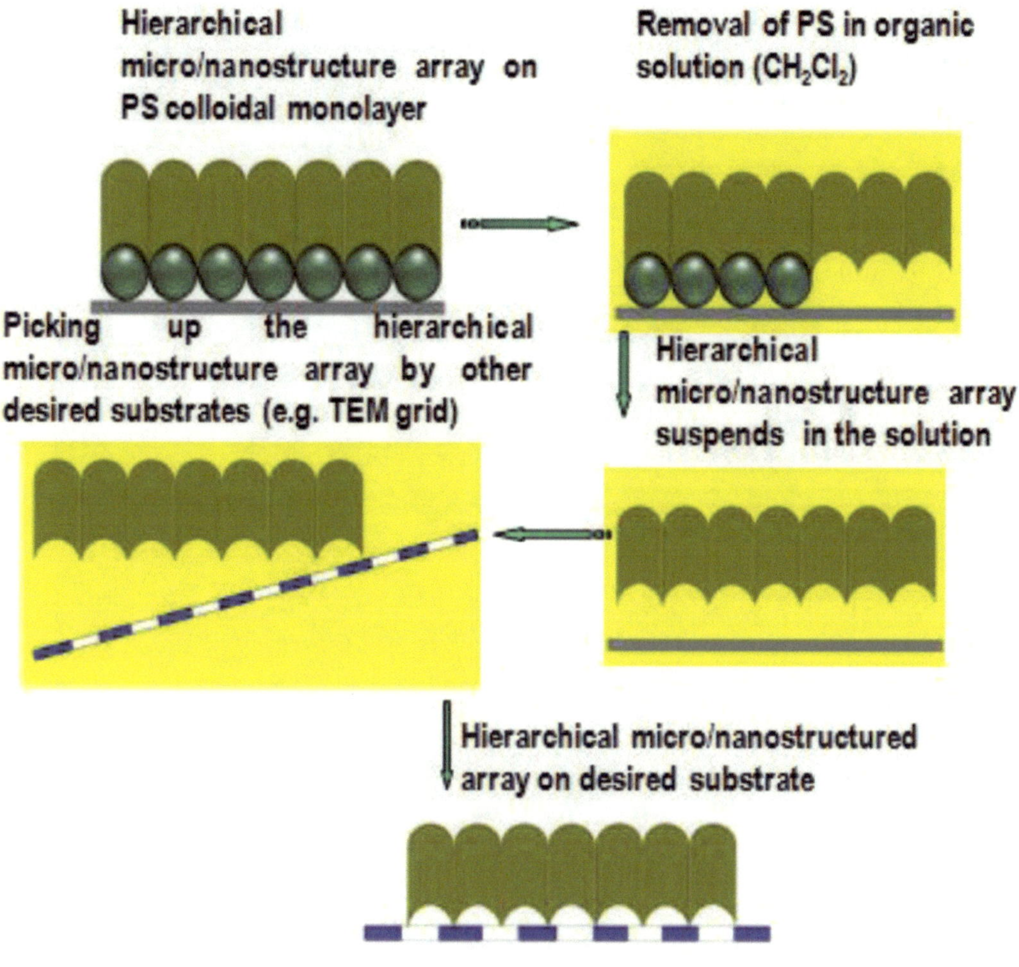

Fig. 9. Schematic illustration of transferability of hcp hierarchical micro/nanostructured arrays

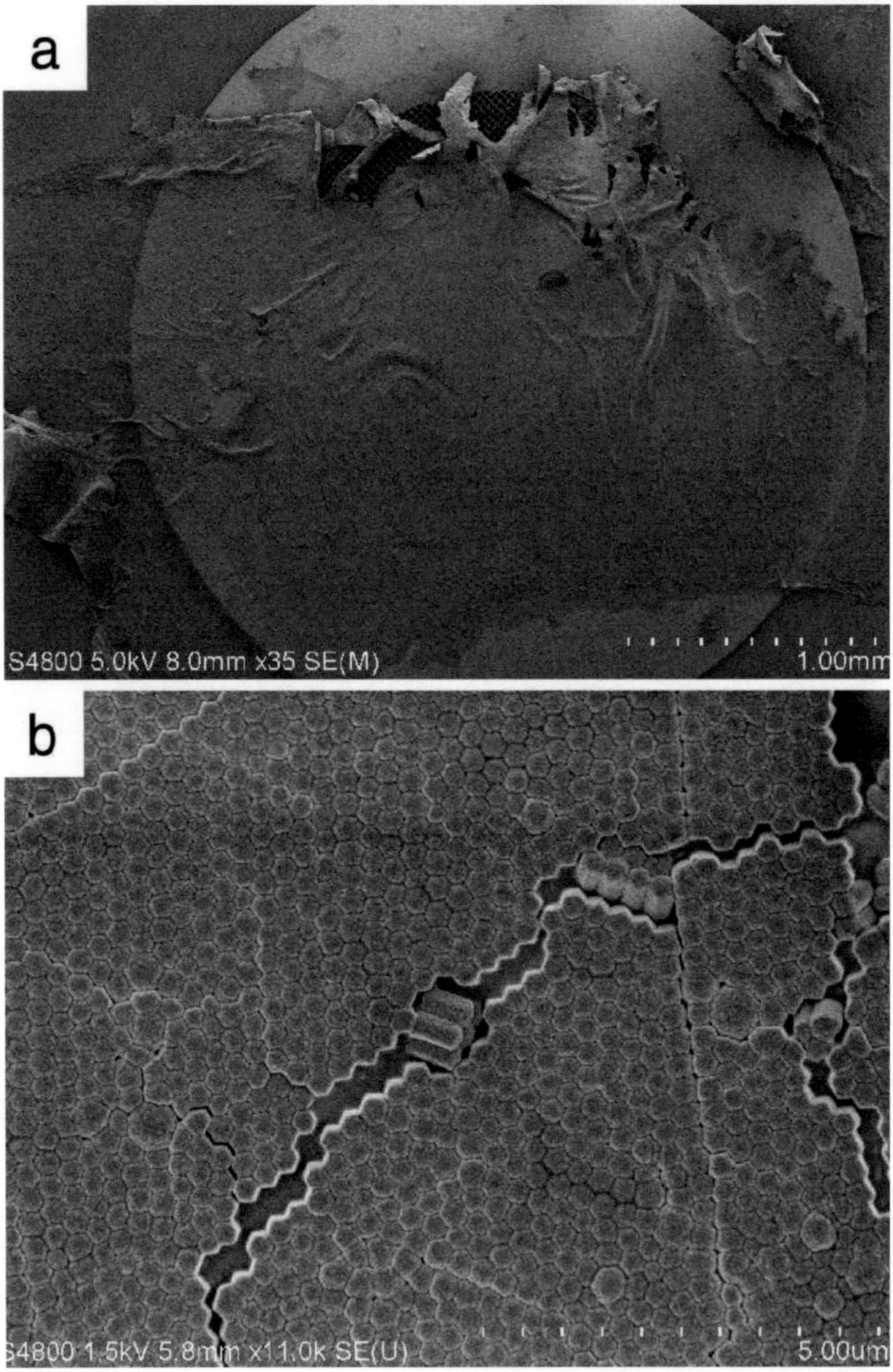

Fig. 10. FE-SEM images of transferred TiO_2 micro/nanostructured arrays from a silicon substrate on a TEM grid. (a) Low- and (b) high-magnification images of array film on a TEM grid.

2.2 Formation mechanism of hierarchical micro/nanostructured arrays by PLD

The formation process was traced by PLD using colloidal clusters with different PS spheres as templates. Herein, the TiO_2 was selected as desired material and colloidal monolayer with PS sphere size of 350 nm as template to demonstrate the formation process of hierarchical micro/nanostructured arrays. The colloidal clusters with different PS sphere were fabricated by spin-coating with a higher rotation speed (2000 rpm) and lower concentration (1.0 wt%) of PS colloidal microsphere suspension. For example, a single PS sphere or PS sphere clusters with different sphere numbers (2, 3, 4,...) can be easily created on the substrate by above route, as indicated in column A of Table 1. After PLD, morphologies observed from the top compared to those before PLD, as demonstrated in column B of Table 1. For a single PS microsphere, the shape kept spherical but that the size increased from 350 nm (PS sphere size) to 500 nm after PLD. For the PS sphere-clusters with sphere number from two to six, each unit size in the sphere-cluster still increased, but could not maintain the spherical shape after PLD. Growth of deposited TiO_2 was restricted at the contact point of two neighboring PS spheres, the contact between the neighboring units changed from a quasidot contact to a facet contact before (PS sphere-cluster) and after (PS sphere-cluster with deposited materials on the surface) PLD. If a PS sphere in sphere-cluster was completely surrounded by others, e.g., the central sphere in a hexagonal close packed (hcp) sphere-cluster of seven, its size after deposition was almost the same as before PLD and the morphology was slightly changed from spherical shape to hexagonal one. A section of a PS sphere-cluster of 10 spheres with hcp arrangement after PLD displays that hierarchical micro/nanorods have formed on the two spheres completely surrounded by the others and that hierarchical rod cannot be formed on the spheres at the edge of the sphere-cluster. This implies that a hierarchical micro/nanostructured array will be easily produced after PLD if a colloidal monolayer with a large-scale is applied in the PLD process. Additionally, if the desired materials are deposited on a bare silicon substrate without any PS spheres by PLD, nanocolumns grow vertically on the Si substrate, as seen in Figure 11.

Generally, nanocolumns prefer to grow in the normal direction on the substrate during the PLD process.[75] In the PLD process, the desired target (TiO_2) is irradiated by a laser beam using an energy level exceeding its threshold in vacuum environment, plasma including ions (Ti^{4+}, O^{2-}, etc.), molecules, electrons and clusters are released into the PLD chamber from the target. However, if a background gas with high pressure is introduced into the chamber, the movement direction of ions or electrons will be changed from an almost uniform direction to multidirection due to collisions between the ions, electrons, molecules and clusters of the ejected species and the background gas. According to the above facts, the formation mechanism of hierarchical hcp nanocolumn arrays can be easily understood, as displayed in Figure 12. If a substrate without PS spheres is used in the PLD process, a film consisting of vertical nanocolumns of small diameter will be formed. If a single PS sphere exists on the substrate, a composite of a PS sphere at the bottom and a shell composed of TiO_2 radiation-shaped nanobranches on sphere top will turn up, due to preferential vertical growth along the normal direction of the supporting surface and multidirectional deposition. For a PS sphere cluster (more than one sphere) on the substrate, a shadow effect will be produced in the deposition between any two neighboring spheres. If one sphere in the sphere-cluster is completely surrounded by six other spheres as in the case of hcp arrangement, one rod with hierarchical micro/nanostructure will grow on this sphere top. If a colloidal monolayer with a large scale is adopted, this route can easily fabricate an hcp hierarchical micro/nanostructured array. In this strategy, an off-axis configuration is adopted where the target and substrate are perpendicularly placed. It is similar to the glancing angle deposition (GLAD) or oblique angle deposition in which there is a large angle between the deposition direction and the normal direction of the substrate.[76-84] In the traditional GLAD method, atoms from the target obliquely arrive and condense on the substrate, and the tilted and separated nanowire or nanopillar array with a porous structure are gradually produced due to the shadow effect of the initial deposited nanoparticles under high-vacuum conditions. The critical difference between this route and GLAD is the background gas pressure during deposition, which converts the directional flow of ejected species in a vacuum into a multidirectional one at higher pressure. Therefore, this multidirection deposition and shadow effect are a principal reason why a vertical hierarchical micro/nanostructured array with hcp alignment is formed on the colloidal monolayer. This can be further verified by varying the angles between substrates and target in PLD process, as seen in Figure 13 and 14. If these experiments were carried out in a vacuum, tilted rod-like structured arrays with different angles would be obtained on the different substrates. However, from these results, the rod-like morphologies are independent of the angle between the substrate and target but the growth rates are different for different angles because of the plume shape in PLD , they always grow vertically on the substrate due the multiple direction deposition combined with shadow effect of neighboring colloidal sphere (Figure 14).

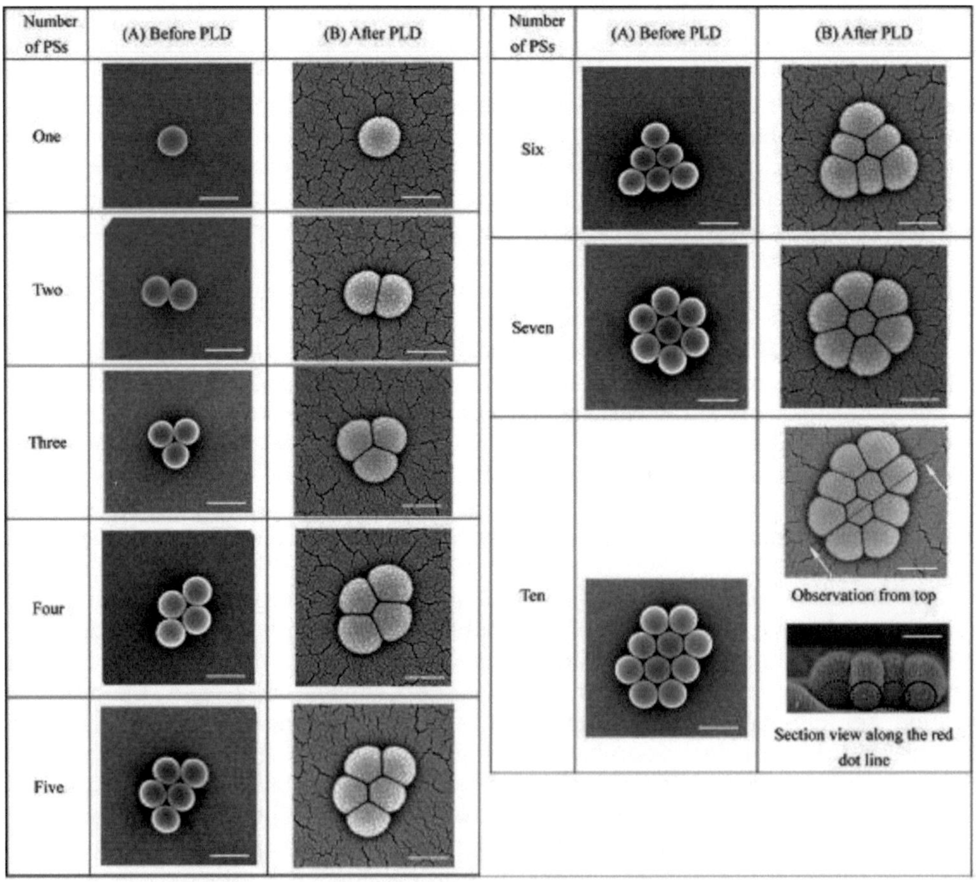

Table 1. Morphologies of before and after PLD on the PS sphere surface (Scale bars are 500 nm)

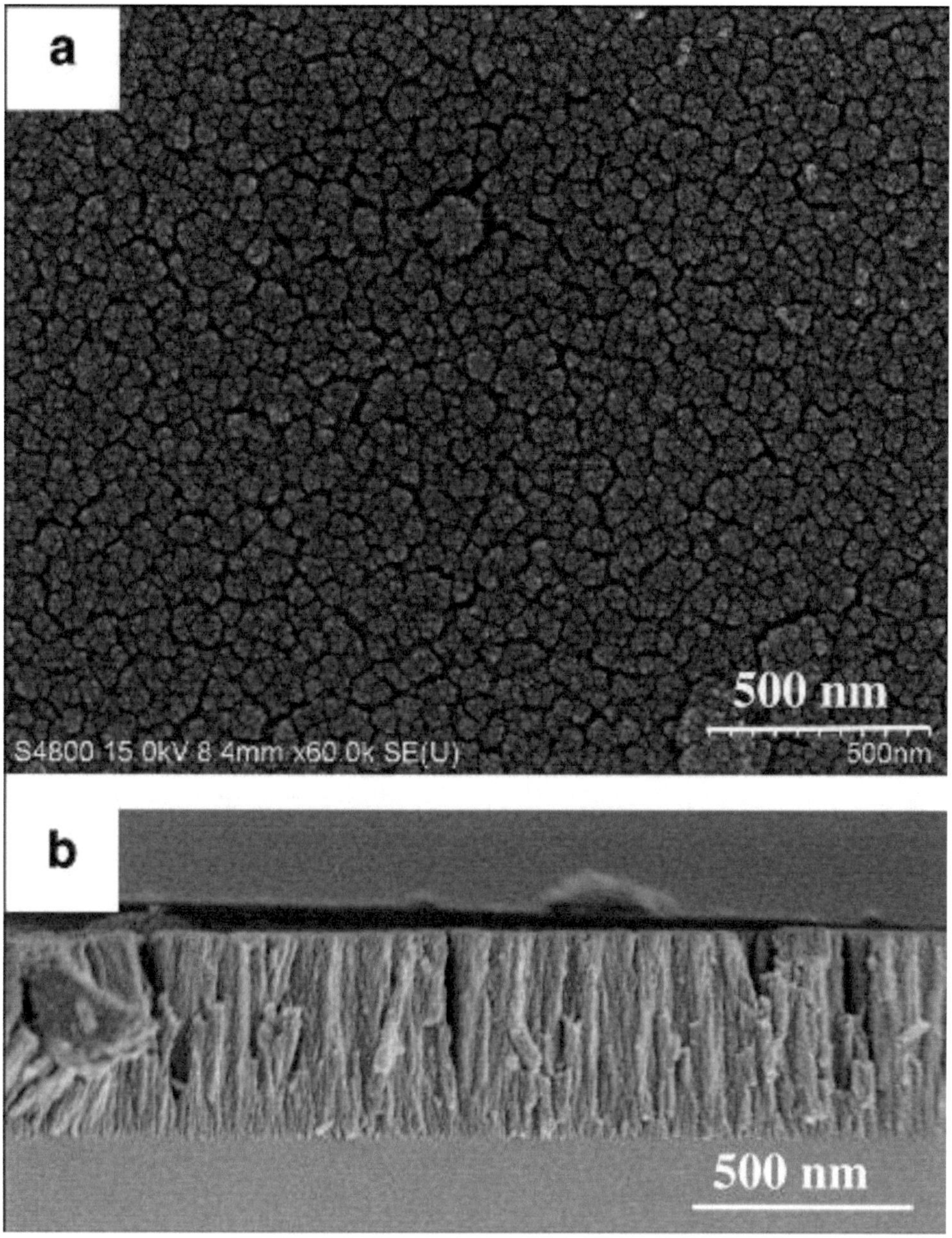

Fig. 11. FE-SEM images of the TiO_2 nanocolumns deposited by PLD directly on a bare substrate without colloidal spheres. (a) Observation from top. (b) Section view.

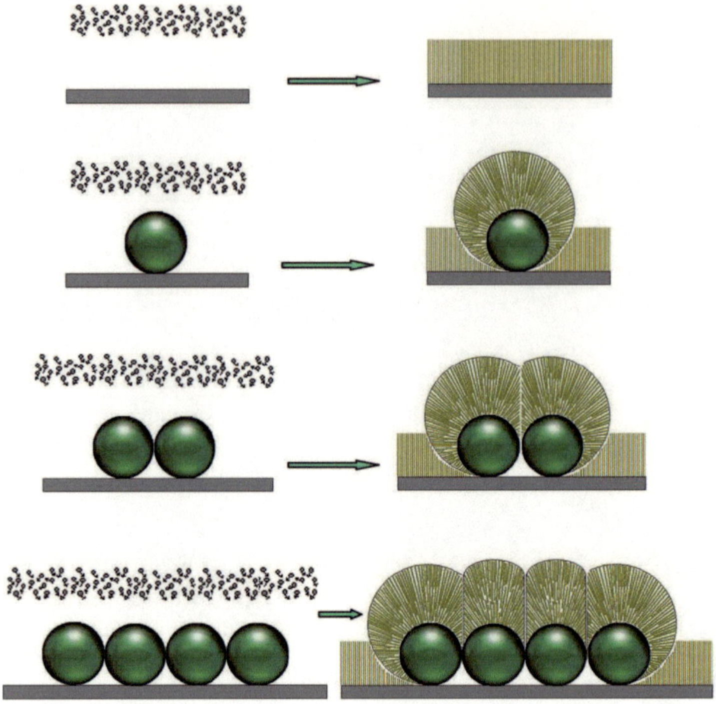

Fig. 12. Schematic illustration of formation mechanism of hcp hierarchical micro/nanostructured arrays.

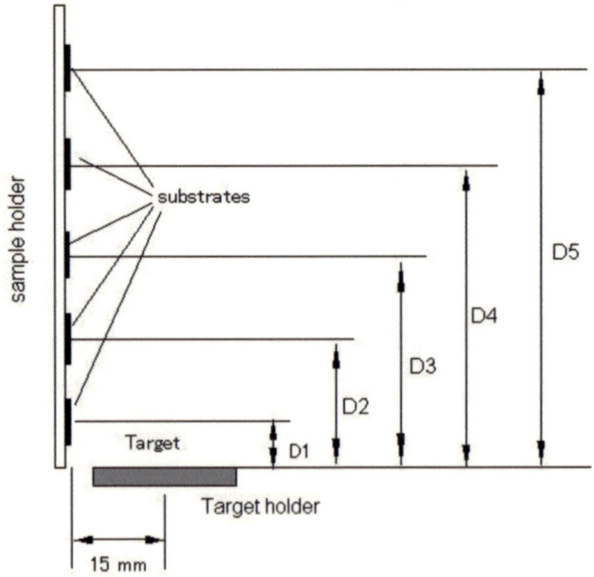

Fig. 13. Schematic illustration of multi-substrate experiment. D1: 8 mm. D2: 23 mm. D3: 35 mm. D4: 50 mm. D5: 64 mm. In this experiment, substrate rotation: 0 rpm. PS sphere size in colloidal monolayer: 350 nm.

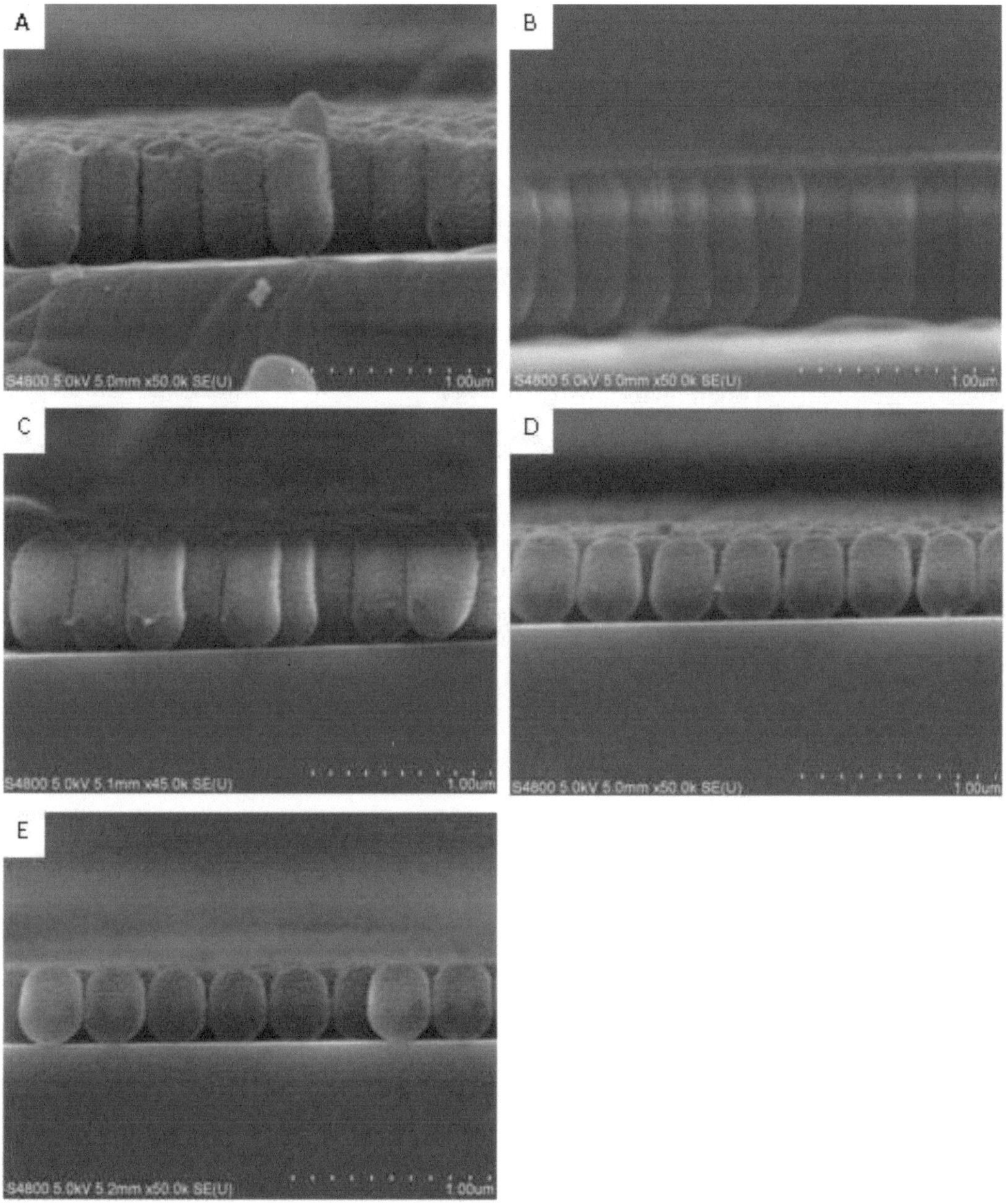

Fig. 14. FESEM cross-section images of the samples at different positions in Figure 10. Images A to E are from the samples at positions D1 to D5.

2.3 Extension of PLD assisted colloidal lithography

The hexagonal close-packed (hcp) micro/nanostructured arrays can be obtained by PLD using colloidal monolayer template. Such PLD assisted colloidal lithography can be extended to prepare hexagonal non-close-packed (hncp) micro/nano- particle or nanorod arrays by further annealing process. [85-88]

PS colloidal monolayer was first fabricated on Si substrates by a self-assembly process. It was then placed in a PLD chamber for deposition at room temperature. After deposition, the sample was moved to an oven from the PLD chamber and annealed in air. The hncp, hierarchically ordered micro/nano- particle arrays were thus prepared on the substrate, as illustrated in Figure 15.[85]

Figure 16 shows FE-SEM images of the TiO_2 hierarchical hncp micro/nano- particle array obtained by PLD assisted colloidal lithography and subsequently annealed at 650 °C for 2 h in air. One can clearly find that this particle-ordered array takes on hncp arrangement. Each particle in the periodic array exhibits a hemispherical shape with an average size of 240 nm (Figure 16c) and is composed of small nanoparticles (Figure 16d). Additionally, SEAD pattern (inset in Figure 16d.) indicates the deposited materials are changed from amorphous to anatase typed TiO_2 after annealing. These results reflect that this hncp particle array possesses a hierarchical micro/nano-structure.

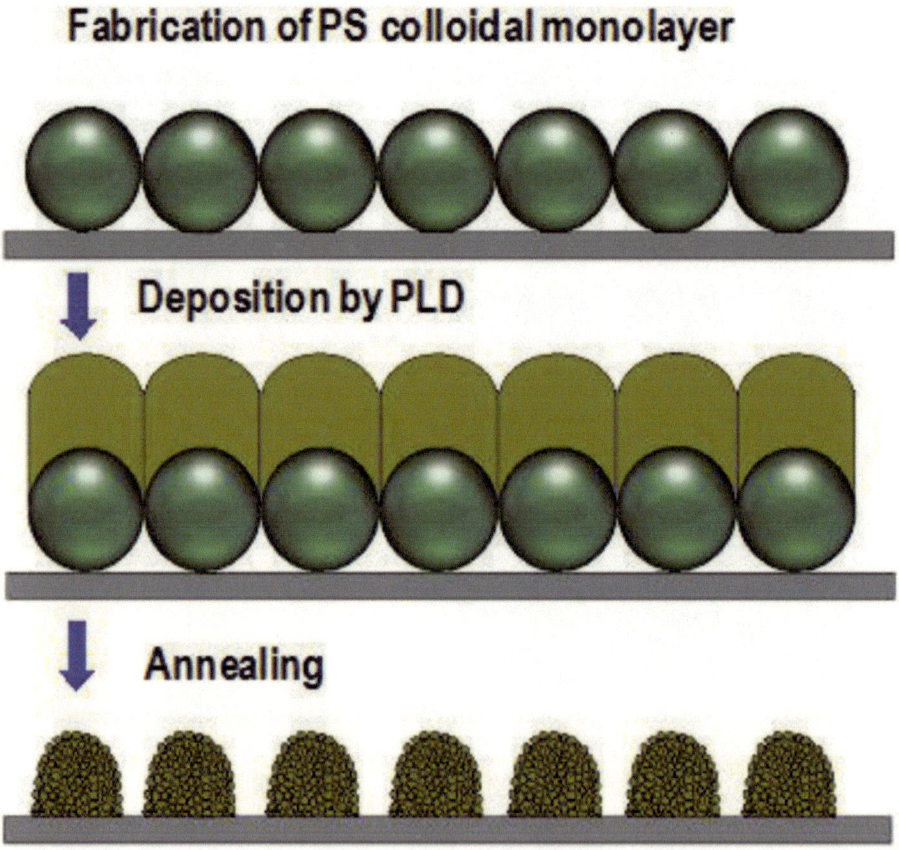

Fig. 15. Schematic illustration of fabrication process for the hierarchical hncp micro/nano-particle array.

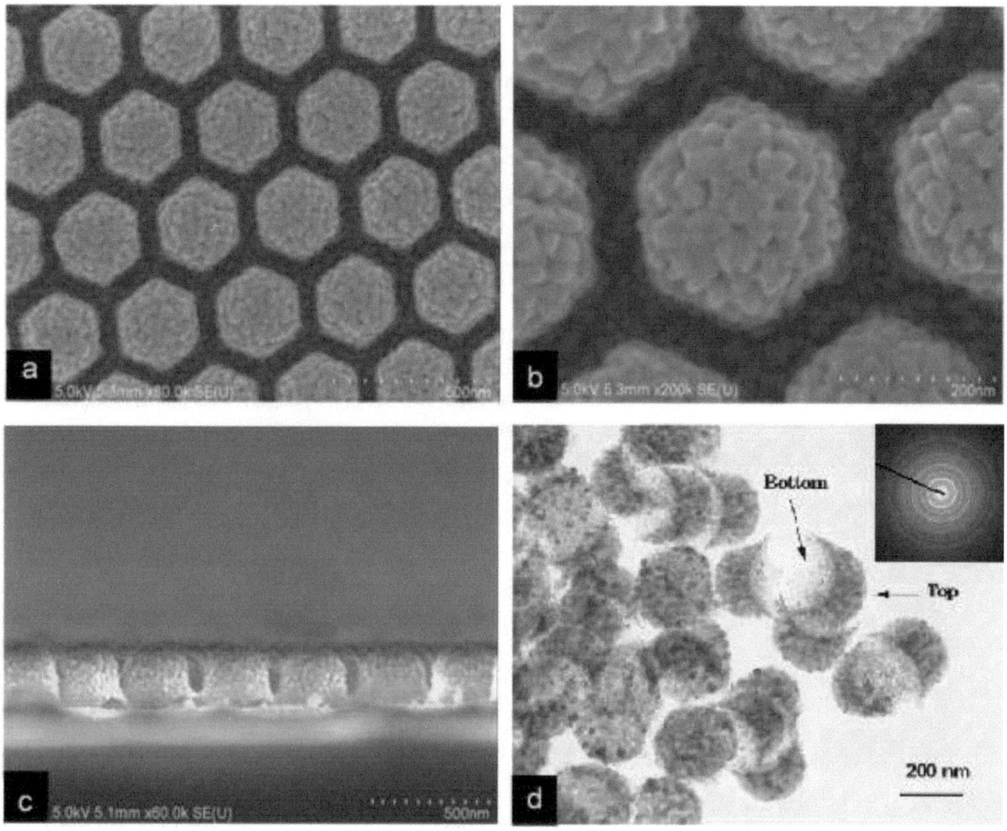

Fig. 16. FE-SEM and TEM images of the hierarchical hncp micro/nano particle arrays. a)-c): FE-SEM images. a) Large area hierarchical micro/nanostructured array b) Same at high magnification. c) Cross section d) TEM image of hierarchical particles (scratched from the supporting substrate by a knife and transferred to TEM copper grid for observation) and the corresponding SAED pattern (inset).

The formation of hncp, hierarchical micro/nano- particle array is discussed based on experimental results. Herein the explanation is given using TiO_2 as an example. The sample produced by PLD assisted colloidal lithography without further heating displayed an hcp alignment (Figure 17a). TiO_2 was deposited on the PS sphere surfaces and grew along the vertical direction (Figure 17b). Each particle is composed of two parts: the PS sphere at the bottom and an amorphous, porous TiO_2 layer consisting of smaller particles on the top (Figure 17c, d). The amorphous materials crystallize after being annealed at high temperature. In this case, when the amorphous TiO_2 with its supporting PS spheres was heated at 650 ^{o}C for 2 h, the PS spheres were burned out. Meanwhile, the TiO_2 particles on top of the PS sphere were changed to anatase polycrystals composed of smaller nanoparticles of ca.30 nm and were dropped vertically down to the original position of the PS sphere. Additionally, the volume of TiO_2 particles decreased during the change from the amorphous to the crystalline phase and hence an hncp hierarchical particle array was formed on the substrate, as illustrated in Figure 18. The hierarchical hncp mciro/nano-particle array film adhered tightly to the substrate after annealing and could not be detached from the supporting substrate even when it was washed ultrasonically in water for 30 min.

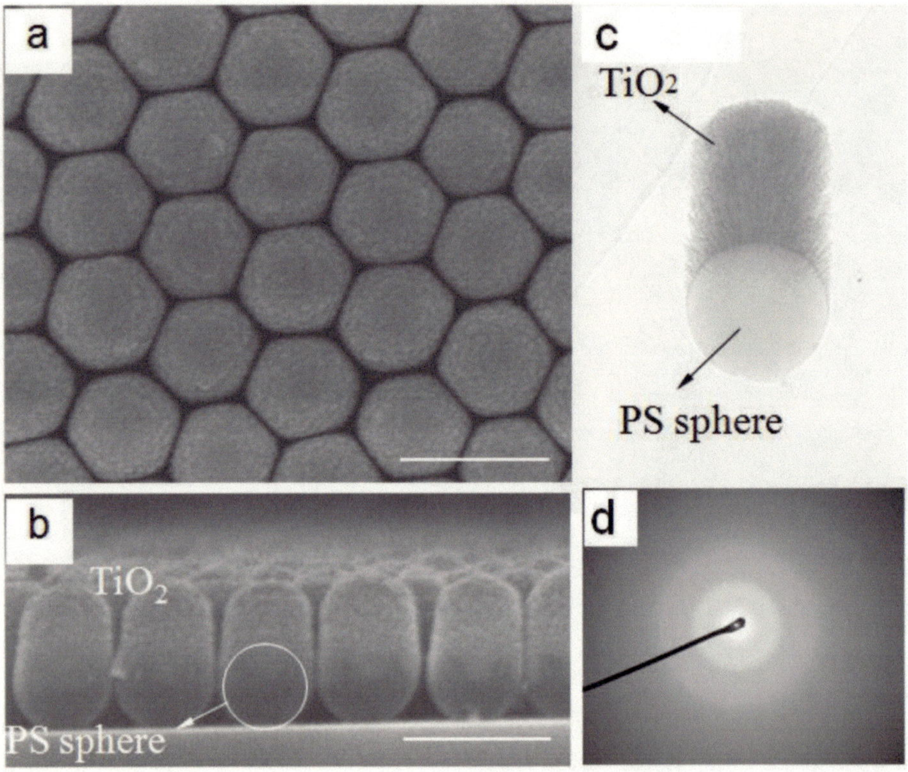

Fig. 17. (a, b): FE-SEM images of the as-deposited sample produced by the PLD in ambient atmosphere without heating. a) Top-view image. b) Cross-sectional image. c) TEM image of single unit. d) SAED pattern of deposited materials.

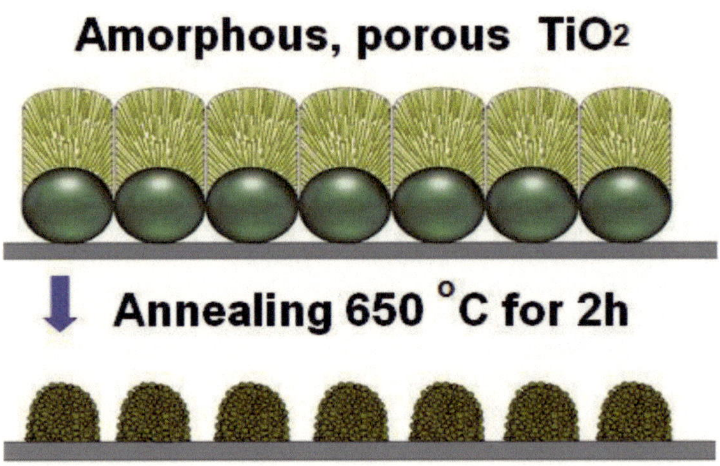

Fig. 18. Formation of hierarchical *hncp* micro/nanostructured particle arrays.

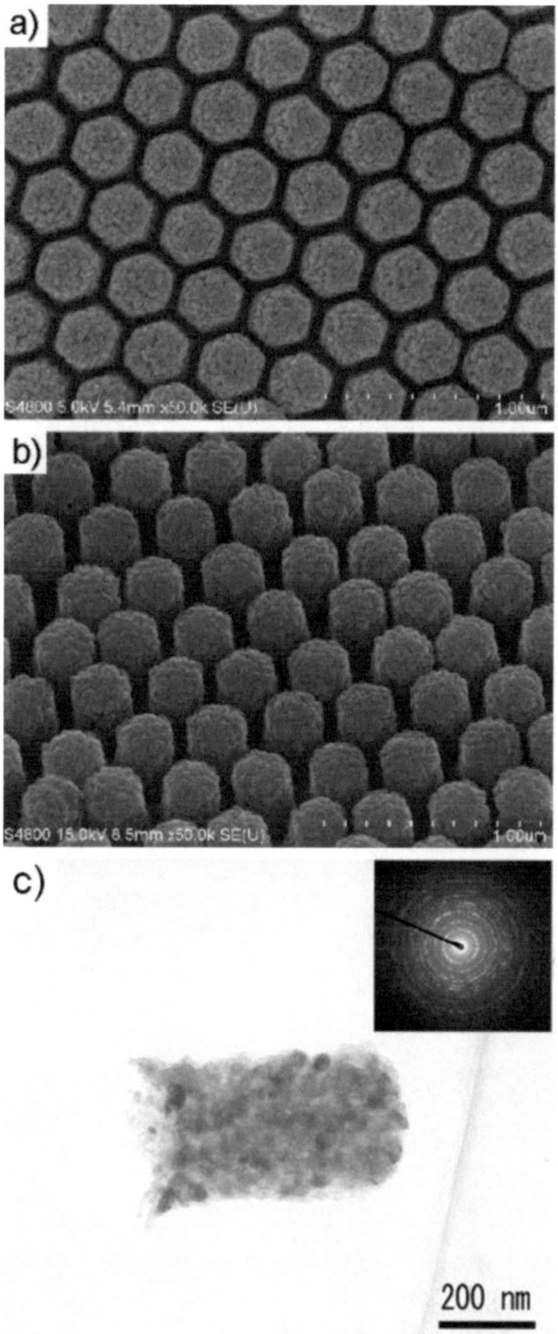

Fig. 19. (a, b) FE-SEM images of a nanorod array by PLD using a colloidal template monolayer with 350 nm PS spheres in O_2 at a pressure of 6.7 Pa for 60 min and subsequent annealing in air. a) A top-view FE-SEM image. (b) An FE-SEM image of a sample tilted to 45o. (c) TEM image of a nanorod and the corresponding selective area electron diffraction (SAED) pattern (inset).

In the PLD assisted colloidal lithography, if the deposition time increases to a appropriate time, an hncp micro/nano rod array can be obtained after PLD and subsequently annealing.[86] Figure 19 presents FE-SEM images of the hierarchical micro/nano rod array obtained by PLD assisted colloidal lithography (PS sphere size: 350 nm; background O_2 pressure: 6.7 Pa; longer deposition time: 60 min) after annealed at 650 °C for 2 h in air. Figure 19 a and b indicate that a periodic nanorod array takes an hncp arrangement and that each nanorod consists of many nanoparticles. The TEM image (Figure 19c) of a single nanorod shows that it has an aspect ratio of ca. 2:1 and is composed of small nanoparticles, and the PS sphere templates were entirely removed during annealing. The hncp micro/nano- rod arrays originated from amorphous, hcp micro/nanostructured array produced by PLD at room temperature without annealing, as displayed in Figure 20. Beside amorphous materials deposited by PLD, some crystalline materials, e.g. Fe_2O_3, their hncp micro/nanostructured arrays can be also achieved by PLD assisted colloidal lithography after annealing. Figure. 21a, b present SEM images of hierarchical, hcp, crystalline Fe_2O_3 micro/nanostructured arrays by PLD at oxygen pressures of 6 Pa at room temperature. Such sample was annealed in air at 450 °C for 3 h and, PS colloidal monolayer template was completely decomposed and hierarchical hncp micro/nano-structured array was formed, as shown in Figure 21 c-f.

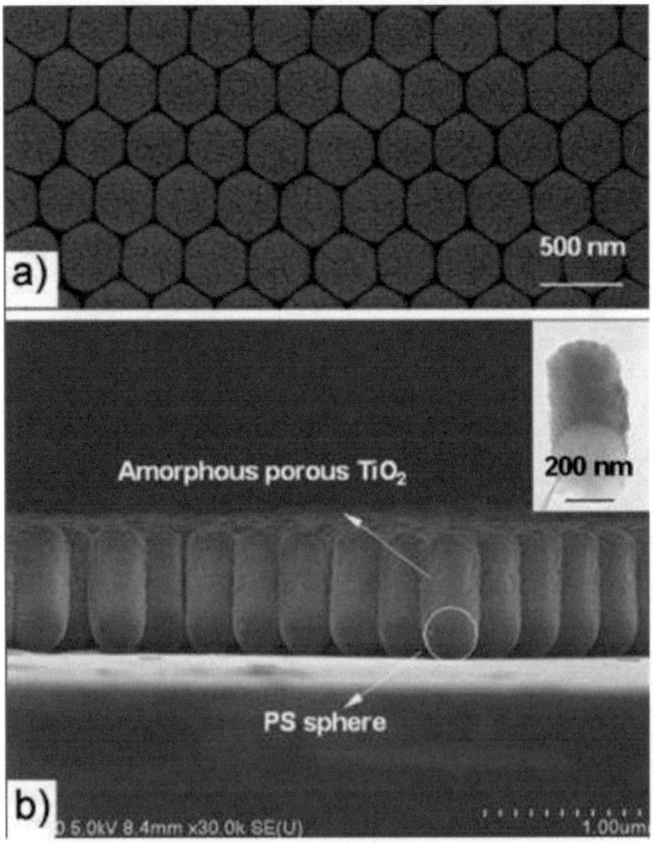

Fig. 20. FE-SEM images of an amorphous TiO_2 hcp nanorod array on the colloidal monolayer before annealing. a) Observation from top, b) Cross section. Inset in (b): a TEM image of a single amorphous TiO_2 nanorod with PS sphere.

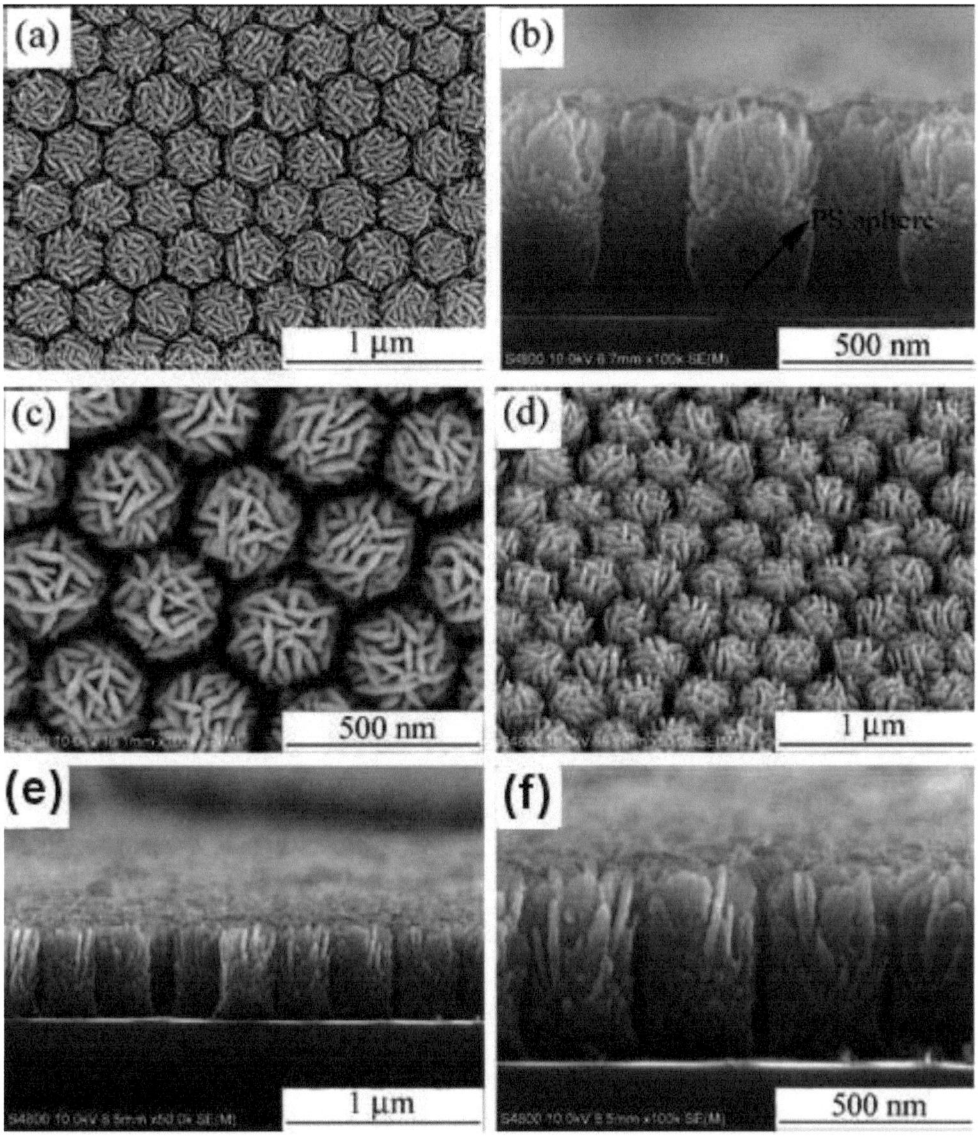

Fig. 21. SEM images of Fe_2O_3 micro/nanostructured array based on colloidal monolayer template and PLD at the oxygen pressure of 6 Pa. (a, b) Before (c-f) after annealing 450 °C for 3 h in air. (c) Top-surface image. (d) 45° tilted view. (e) Cross-section image. (f) Higher magnification.

In this route, the interspace between two neighboring rods can be controlled by changing the background gas pressures in PLD process. Usually, the porosity and specific surface area of the rods in arrays can be tuned by varying the background gas pressure in PLD deposition and they increase with increase of background gas pressure (Figure 22a–c). Therefore, the interspace between neighboring anatase rods can be tuned with different background gas pressures in an hncp array after annealing. The interspace will increase with increasing background gas pressure (Figure. 22a′–c′ and a″–c″).

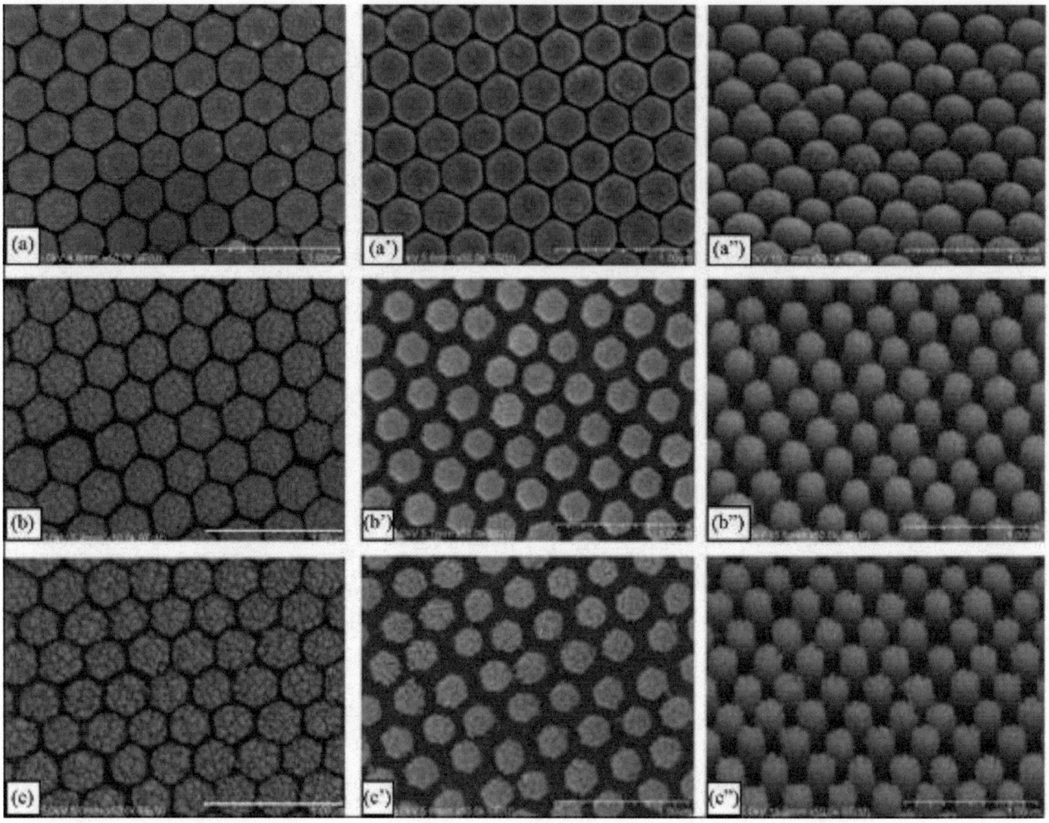

Fig. 22. FE-SEM images of an amorphous hcp TiO_2 nanorod array and anatase hncp TiO_2 nanorod arrays obtained by PLD under different background gas pressures and subsequent annealing. PLD was performed in oxygen (a) at 2.0 Pa for 200 min; (b) at 16.8 Pa for 43 min; (c) at 26.8 Pa for 30 min. (a–c) before annealing, (a′–c′) top views after annealing, (a′′–c′′) tilted view with 45 degrees after annealing.

3. Sputtering assisted colloidal lithography

In this method, a PS colloidal monolayer on a substrate is placed into the chamber of radio frequency (RF) magnetron sputtering for material deposition at room temperature and Ar is introduced as the background gas.[89] A unique hncp hierarchical micro/nanostructured array is formed due to PS-templated plasma etching/deposition in a relatively high vacuum (0.06 Pa). The features of this technique are (1) low-pressure sputtering; (2) PS-templated sputtering, which guarantees a periodic arrangement, and (3) plasma etching/deposition, which eventually produces the unique hncp hierarchical micro/nano-structure.

Figure 23 presents FE-SEM and TEM images of hierarchical alumina mciro/nanostructured arrays by sputtering target alumina using a PS colloidal template with a sphere size of 750 nm in a relatively high vacuum (0.06 Pa). They show the following three unique features. First, it is a periodic nanocolumn array cushioned by a semi-shell in an hncp arrangement. Each nanocolumn is composed of two parts: a semishell- shaped cushion at the bottom and a vertical nanocolumn on the top of the cushion. Such nanocolumn possesses a very rough structure on the surface and seems to be composed of many minicolumns, indicating that the sample possesses a hierarchical, porous structure and hence has a high surface area. Secondly, the periodicity was 750 nm, matching well with the initial size of the PS spheres. It is very evident that the sizes of the cushion and the central columns were reduced by about 15% and 45% compared to the original size of the PS template.

The formation of such hierarchical micro/nanostructured arrays is traced by the different sputtering deposition time, as demonstrate in Figure 24. With increase of deposition time (Figure 24 A–D), the PS sphere size gradually decreases and the alumina columns grow vertically in the center, finally the columnar structures and a salver-shaped semi-shell are formed. Generally, thin alumina continuous film is formed on bare substrate without PS spheres due to the strong ion energy and subsequent rapid surface migration under such a low sputtering pressure. In the case with PS sphere array on the substrate, alumina components sputtered from the target are impinged and implanted into the PS due to the strong ion energy and soft nature of PS. The part of PS sphere is also continuously etched away by argon ions and part of deposited alumina is also etched away, but remaining part will gradually form a structure. Additionally, the PS colloidal monolayer supplies the periodic array template. Merging these two aspects into one forms a unique hierarchical micro/nano-structured arrays. The PS spheres become smaller with plasma etching and a salver-shaped semi-shell gradually appear. Further sputtering causes the PS spheres to be etched more significantly, and the species generated from the target deposit perpendicularly onto the template (both the center and the semi-shell part), thus forming a column structure and salver-shaped semi-shell. Implanted components of aluminium and oxygen into PS sphere will be linked together by continuous etching of PS. But at the side edge of the spheres, the amount of PS is not much and easily etched away to form aluminium oxide film, resulted in the cushion shell. The amount of PS at the center part is much more and even by continuous etching a film cannot be formed and rod-like structures are generated. Further sputtering continues etching the PS sphere until the final unique hncp hierarchical structure forms. The formation process of this unique hierarchical mciro/nano-structure is schematically illustrated in Figure 25. In order to further confirm this process, the pressures of Ar were adjusted from low level (0.06 Pa) to 0.13 and final 6.7 Pa. The FE-SEM images of the samples are presented in Figure 26. With increase of the background gas pressure from 0.06 Pa to 0.13 and 6.7 Pa, the collision probability between the ejected species and Ar molecules increases, thus the PS spheres are more significantly etched and no semi-shell can form. Therefore, only columnar structures are obtained (Figure 26). The amounts of deposited materials in the inter-columnar structures are negligible probably due to the blocking effects of gaseous species emitted from decomposed PS spheres during sputtering. These results firmly prove that a relatively high vacuum condition subsequently induces mild plasma etching/deposition. Besides hncp alumina micro/nano-structured arrays with a periodicity of 750 nm, novel hierarchical arrays with periodicities of 350 nm, 1 μm, and 2 μm were also created by colloidal monolayers with different PS sphere sizes during sputtering at the same pressure of Ar as in Figure 27. Besides alumina, hierarchical arrays of other materials including Au/Al_2O_3 composite, CuO, and NiO can also be fabricated by the presented one-step plasma etching. Some of the results are presented in Figure 28. In this method, only the inorganic materials can be used as the deposited materials. Otherwise, the deposition cannot be guaranteed because of the subsequent etching.

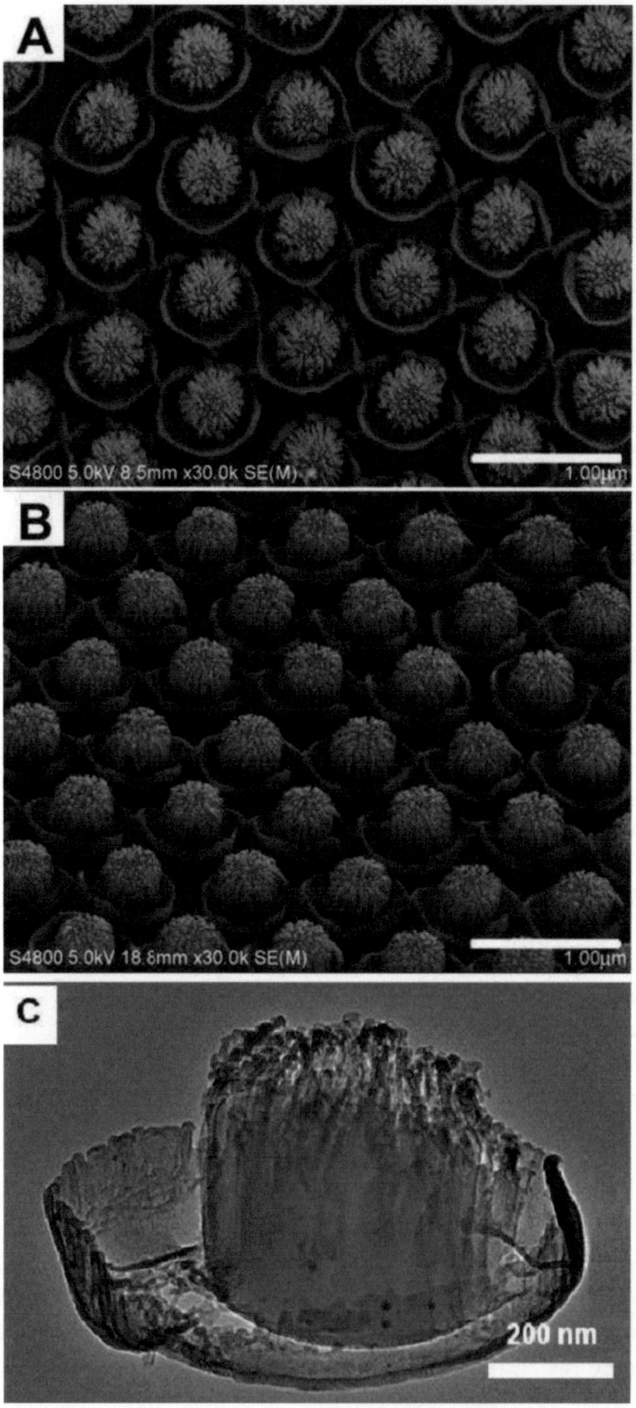

Fig. 23. FE-SEM (A, B) and TEM (C) images of a sample obtained by sputtering using a PS colloidal monolayer as the substrate (PS sphere size: 750 nm; deposition time: 2 h). (A) Top view. (B) Titled view with 45° angle. The scale bar in A, B: 1 μm. (C) TEM image of one unit.

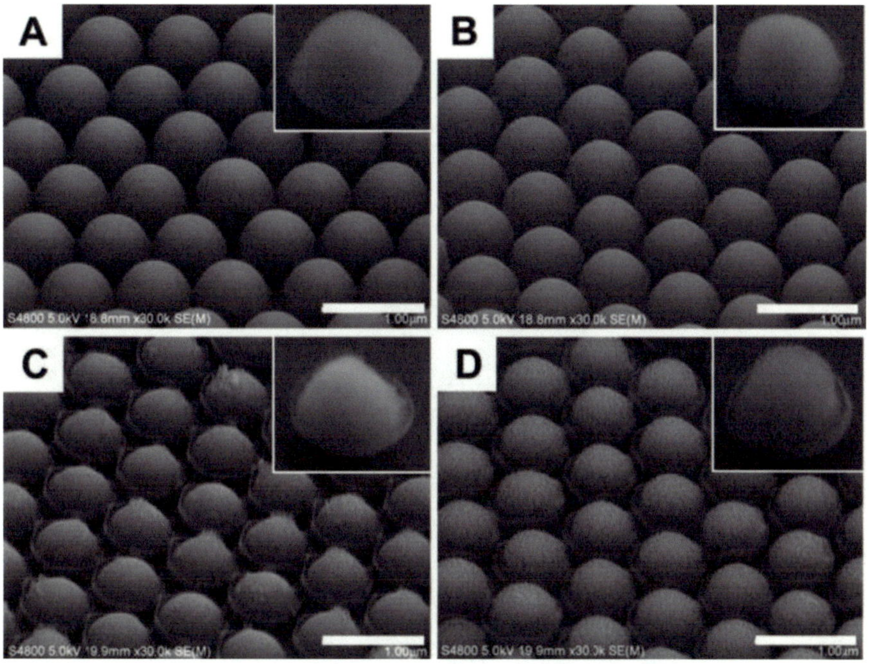

Fig. 24. FE-SEM images of samples obtained with different deposition times. (A) 10 min. (B) 25 min. (C) 30 min. (D) 60 min. The inset is high-magnification image of one unit in each figure.

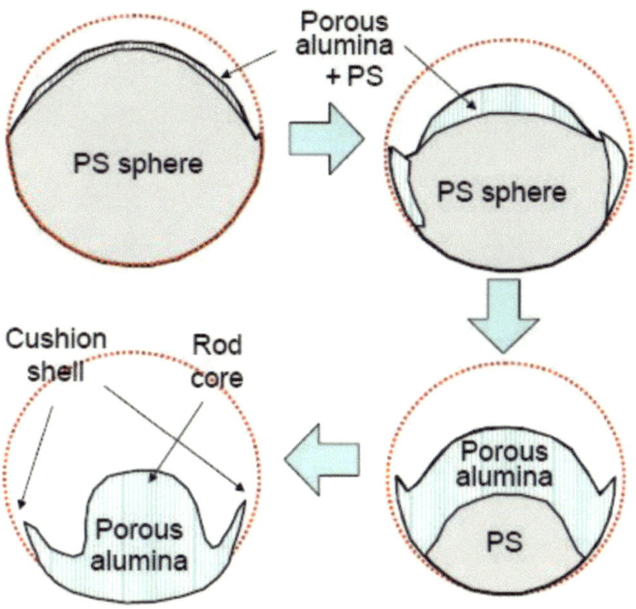

Fig. 25. Schematic illustration of the formation process of unique hncp hierarchical micro/nano-structure.

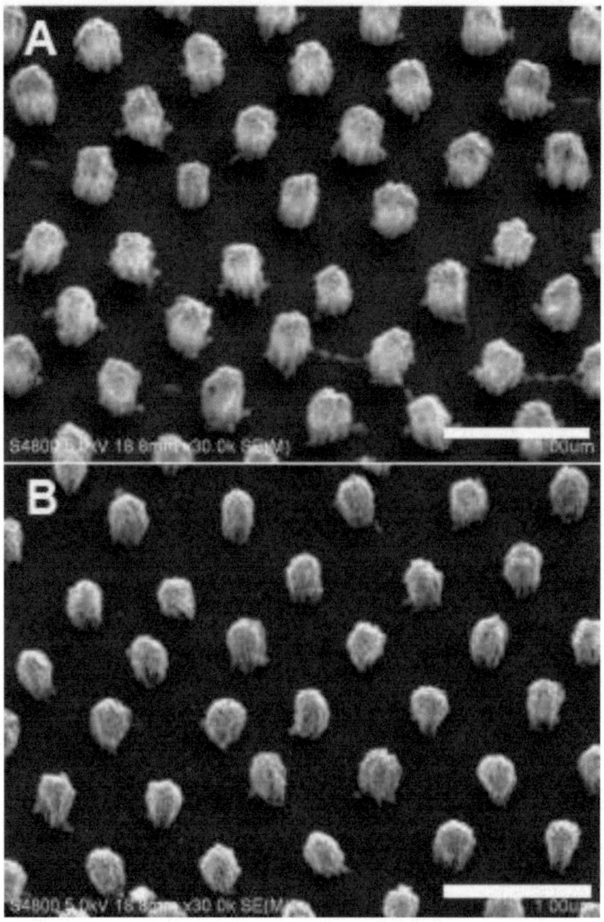

Fig. 26. FESEM images of a sample obtained by sputtering using a PS colloidal monolayer as the substrate at Ar pressures of (A) 0.13 Pa and (B) 6.7 Pa (PS sphere size 750 nm; deposition time 2 h). These images are observed with a tilt angle of 45°. The scale bar is 1 μm.

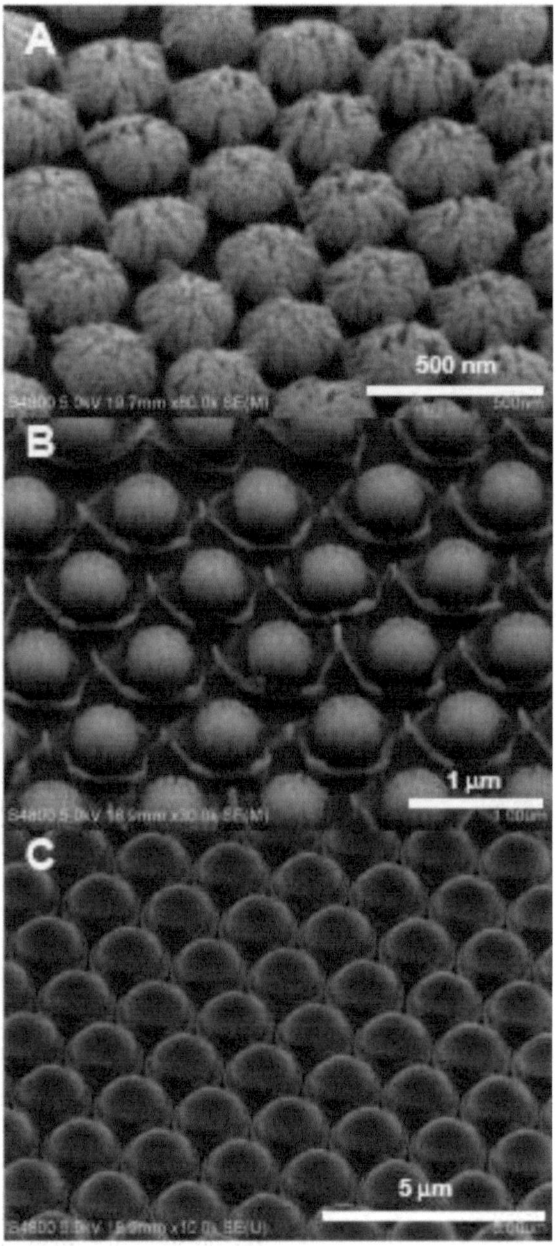

Fig. 27. FESEM images of a sample obtained by sputtering using a PS colloidal monolayer template with different PS sphere sizes at 0.06 Pa. (A) 350 nm. (B) 1 µm. (C) 2 µm.

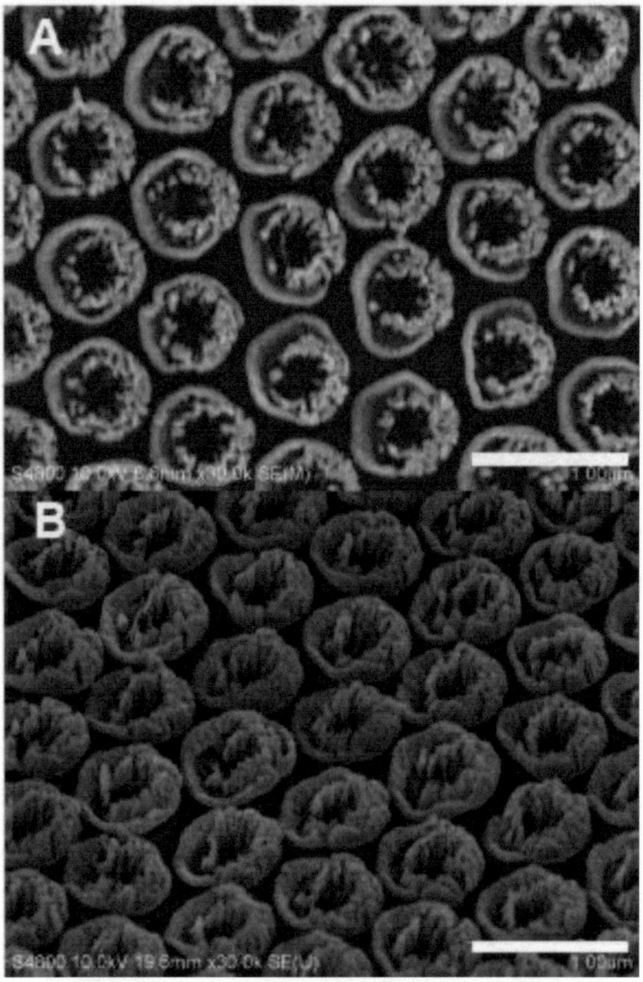

Fig. 28. FESEM images of periodic Au/Al_2O_3 nanocomposite arrays obtained by co-sputtering multiple targets consisting of an Al_2O_3 target and Au sheets and using a PS colloidal monolayer as the substrate (PS sphere size 750 nm; deposition time 2 h). (A) Image observed from the top. (B) Image with a tilt angle of 45°. The scale bar is 1 μm.

4. Applications of periodic micro/nanostructured arrays

Based on colloidal monolayer templates, the different hierarchical micro/nanostructured arrays can be fabricated by physical deposition and their morphologies and structural parameters (sizes and interspace of micro/nanostructured unit) can be facilely controlled by periodicities of colloidal monolayers, experimental conditions such as deposition time, background gas pressure etc. Some properties, such as the surface wettability, field emission, catalytic properties are closely morphology and structural parameter-dependent. These properties can be readily optimized by changing morphologies and parameters of the periodic arrays. Their investigations supply useful theoretic foundations and are highly valuable for designing micro/nanodevices based on these special arrays.

4.1 Wettability

Wettability is generally related to the surface morphologies, roughness and free energy of materials surface and it is evaluated by the water or oil contact angle. A special surface with self-cleaning effect is usually defined as a surface that has the ability to remove dirt or contaminants that are on it when water droplets slide along the surface. Self-cleaning is closely related to surface wettability[90-94]. The self-cleaning effect is normally attributed to superhydrophobicity (water contact angle (CA) exceeding 150° and sliding angle (SA) less than 10°) or superhydrophilicity (water CA less than 10°) of the surface. For superhydrophobicity with a self-cleaning effect, contaminants adhere to the water droplet surface and are removed after the water droplet slides off the solid surface with a small tilted angle, due to large water CA and low surface free energy. For superhydrophilic surfaces, contaminants can easily be swept away by adding water droplets on them, due to very low water CA. Wettability can be enhanced by increasing surface roughness, according to Wenzel's equation: [95]

$$\cos \theta_r = r \cos\theta \tag{1}$$

where r is the roughness factor, defined as the ratio of total surface area to projected area on the horizontal plane; θ_r is the CA of film with a rough surface; and θ is the CA of film with a smooth surface. Obviously, increased roughness can enhance the hydrophobicity and/or hydrophilicity of hydrophobic and/or hydrophilic surfaces. The hierarchical micro/nanostructured arrays based on colloidal monolayers are actually rough films at the micro/nano-scale level. It is expected that such hierarchical micro/nanostructured arrays could induce surface superhydrophilicity or superhydrophobicity with a self-cleaning effect, due to their high roughness. Amorphous, porous hierarchical TiO_2 micro/nanostructured arrays were prepared by PLD assisted colloidal lithography (Figure 3)[73]. These arrays exhibited strong superhydrophilicity. When a small water droplet was dropped on a hierarchical structured array, the droplet spread out rapidly on the surface and displayed a water CA of 0° in a 0.225 s (Figure 29). Additionally, this hierarchical array film exhibited superoleophilicity when a small oil droplet was placed on the nanorod surface and the oil CA became 0° in 0.5 s (Figure 30). These results suggest that this amorphous hierarchical micro/nano-structured array had superamphiphilicity with 0° of both water CA and oil CA. A TiO_2 film with superamphiphilicity can generally be obtained by UV irradiation, due to hydroxyl ions generated by oxygen defects or dangling bonds on its surface, induced by photochemical processes[96]. However, the TiO_2 hierarchical micro/nano- structured array film possessed superamphiphilicity without further UV irradiation. The ions (e.g., Ti^{4+}, and O^{2-}) and electrons are released into the PLD chamber and some oxygen species are lost in the vacuum environment in PLD after a TiO_2 target absorbs energy from laser irradiation by exceeding its threshold. Oxygen vacancies are produced in the deposited TiO_2 during PLD, converting relevant Ti^{4+} sites to Ti^{3+} sites that are favorable for dissociative water adsorption. Therefore, these defect sites microscopically form hydrophilic domains on the TiO_2 surface. However, the other parts surrounding the hydrophilic domain remain oleophilic on the surface. A composite TiO_2 surface having hydrophilic and oleophilic domains on a microscopically distinguishable scale demonstrates macroscopic amphiphilicity on the TiO_2 surface[96]. Additionally, a TiO_2 nanoparticle film prepared by PLD without a PS colloidal monolayer exhibited a water CA of 15° and an oil CA of 27° (Figure 31). The roughness of the hcp TiO_2 hierarchical micro/nano-structured array film was greatly increased compared with that of the nanoparticle TiO_2 film produced by PLD without using a colloidal monolayer. According to Wenzel's equation, wettability is enhanced from amphiphilicity to superamphiphilicity. Therefore, the superamphiphilicity of the amorphous micro/nano-structured array originates from the combination of the amphiphilicity produced by PLD and the special rough structures of hcp hierarchical arrays.

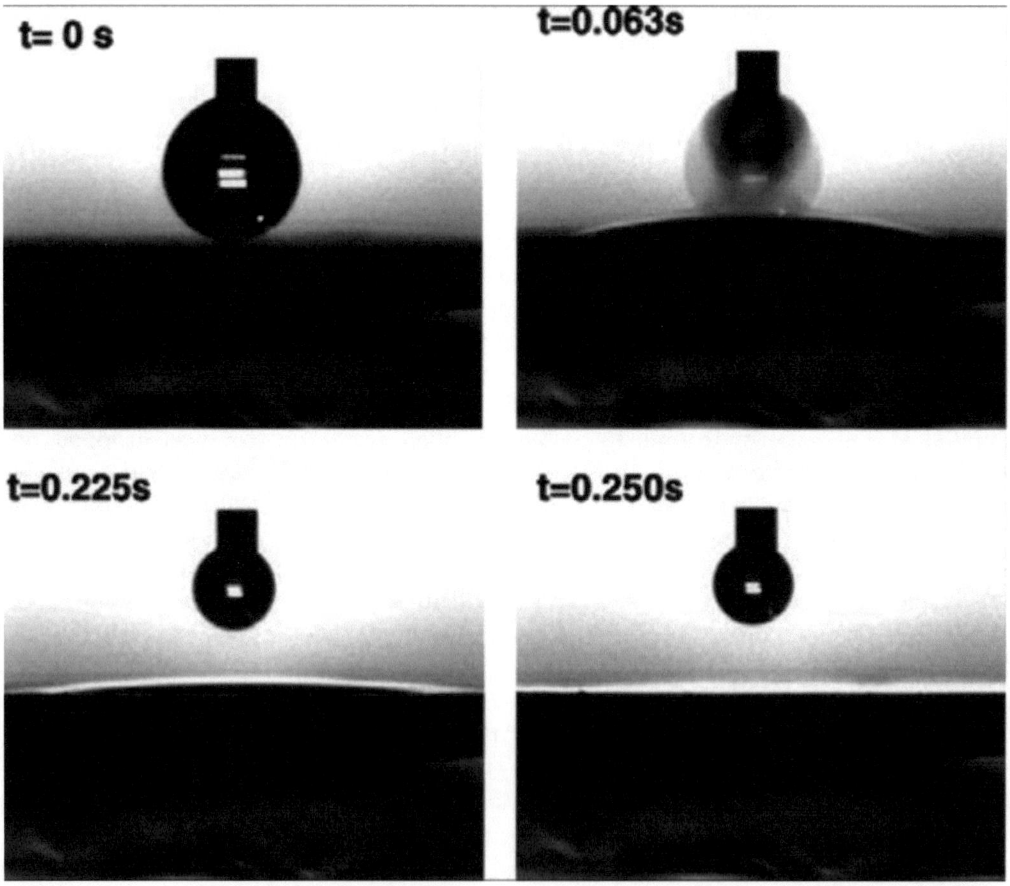

Fig. 29. Time course of water-contacting behavior on the amorphous TiO2 micro/nano-structured array film.

Fig. 30. Oil (rapeseed) droplet shape on amorphous TiO_2 micro/nano- structured array film. The oil contact angle becomes 0° in 0.5 s after it was dropped onto the surface.

More importantly, this amorphous TiO_2 hierarchical array demonstrated very good photocatalytic activity for organic molecular degradation (e.g., effective decomposition of stearic acid under UV illumination). A combination of superamphiphilicity and photocatalytic activity can yield a self-cleaning surface. For instance, an oily liquid contaminant spreads out on a surface due to superoleophilicity, which is helpful in improving the photocatalytic efficiency under light illumination. An organic contaminant including oil gradually degrades under sunlight irradiation (sunlight contains some UV light). The self-cleaning effect can be realized after washing away contamination from the superhydrophilic surface.

Additionally, superhydrophobic surfaces with large water CA and small SA have a self-cleaning effect. For superhydrophobic film, the surface should be sufficiently rough and have a chemical coating with low free-energy materials in order to trap the air on the rough surface. In this case, the area fraction of a water droplet in contact with the sample surface is very small, which helps obtain a small SA. Hierarchical periodic micro/nanostructured arrays based on colloidal templates provide surfaces with regularly ordered and well-defined roughness. They may lead to enhancement from hydrophobicity to superhydrophobicity on the surface after modification with low free-energy materials[97-99]. For instance, Co_3O_4 hierarchical, hncp micro/nano- rod arrays was created by PLD assisted colloidal lithography after annealing at 450 °C for 3 h (oxygen pressures: 93.1 Pa), as shown in Figure 32 a and b.[88] Such surface was chemical modification with fluorosilane, a kind of low free energy material, it presented superhydrophobicity with water CA of 152.6° and a very small SA, indicating self-cleaning effect. It can be explained by Cassie and Baxter equation,[100]

$$\cos\theta_r = f_1 \cos\theta - f_2 \qquad (2)$$

Here, f_1 and f_2 are the surface area fractions of the projecting solid and air ($f_1 + f_2 = 1$). The large fraction of air trapped in the nanorod arrays forms a cushion at the film-water interface that prevents the penetration of water droplets into the grooves. In this case, θ_r is

152.6° and θ is 18.8°, so a value for f_1 of 0.06 is calculated from eqn (2) (i.e. f_2 is 0.94), implying that only 6% of the observed contact area beneath a water droplet is in contact with the water droplet. High f_2 of 0.94 means that the air was well trapped into the groove among nanorod arrays and hence the water droplet kept a spherical shape.

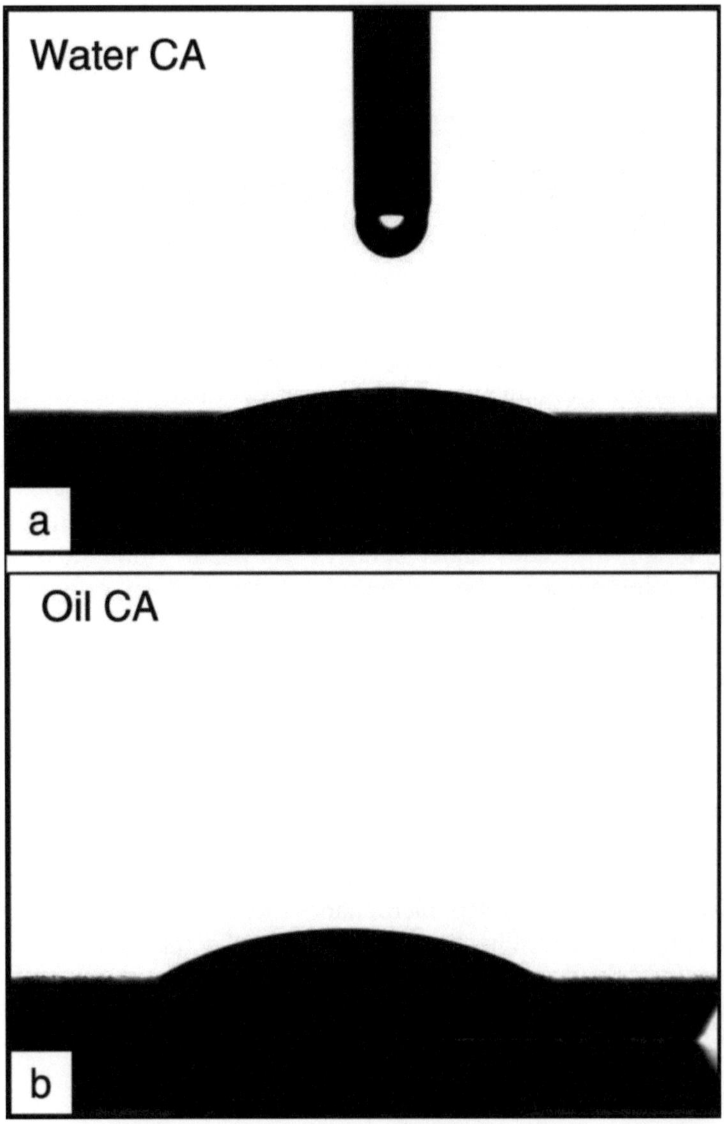

Fig. 31. Water and oil CAs on a TiO_2 film on a silicon wafer prepared by PLD without using a PS colloidal monolayer. (a) Water CA: 15 degrees. (b) Oil (rapeseed) CA: 27 degrees.

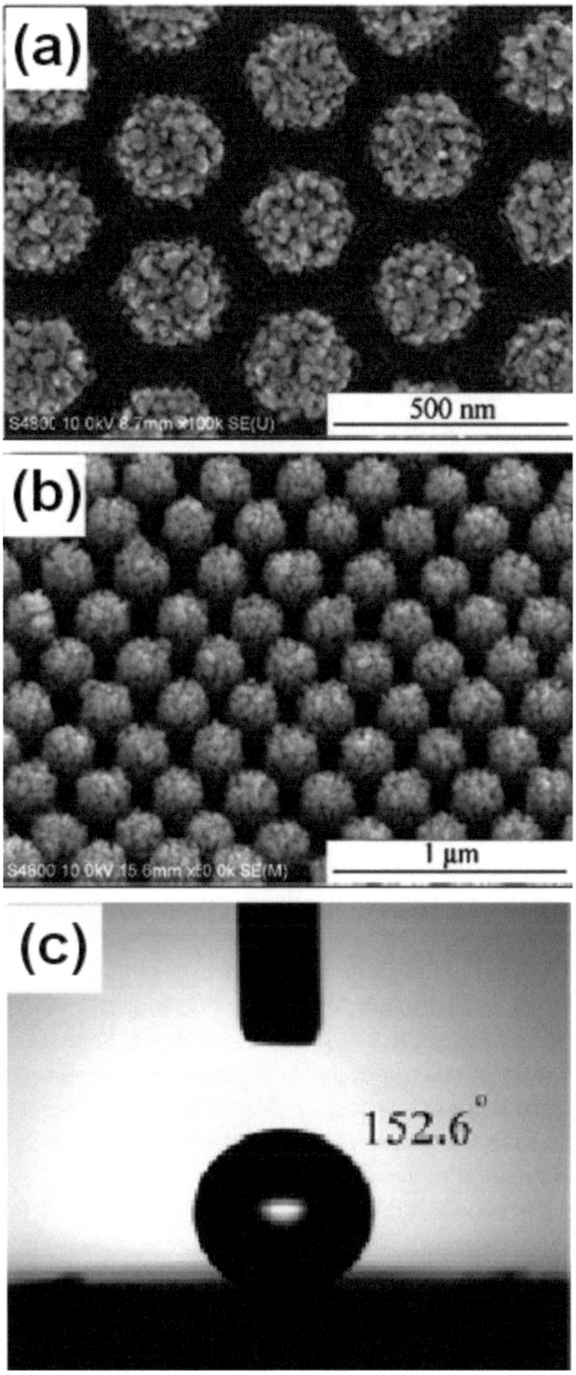

Fig. 32. FE-SEM images of Co_3O_4 hierarchical micro/nano- rod arrays obtained by PLD assisted colloidal lithography after annealing at 450 °C for 3 h (oxygen pressures: 93.1 Pa). (a) Top views; (b) Tilted at 45° (c) water CA after modification with low free energy materials.

4.2 Field emission

Field-emission (FE) properties have recently attracted so much attention due to great commercial interest in flat-panel displays and other microelectronic devices.[101] Besides carbon nanotubes, semiconductors have also attracted great interest in field emitters owing to their good mechanical stability, low work function, and high electrical and thermal conductivities.[102] FE properties are usually decided by the nature of the cathode materials as well as geometry and size of them. By well designing the geometry and size of cathode, for example, to introduce nanostructures on it, good FE performances have been achieved including faster turn-on time, compactness, and sustainability during the field emission compared to conventional bulky material forms. More importantly, researchers have found that cathodes composed of periodic regular arrays on the surface are very highly helpful for producing a low operating voltage and a stable current because of the elimination of the shield effect on densely packed 1D nanostructured arrays in field emission.

Periodic TiO_2 micro/nano-rod arrays with hncp arrangement can be synthesized by combining a colloidal monolayer template with pulsed laser deposition (PLD) followed by annealing in ambient air, as described before.[86] By this route, the periodicity of such special nanorods can be easily tuned by changing the colloidal sphere size in the colloidal monolayer template. While a distance between neighboring nanorods can be controlled by varying the background gas pressure during the PLD process if periodicity is fixed for a nanorod array. The well tunable periodicity and distance between neighboring nanorods are very useful for investigating and optimizing their FE performance.[86]

The periodic hncp TiO_2 nanorod array was fabricated by PLD using a colloidal monolayer template with 350nm PS spheres at 6.7 Pa O_2 for 60 min and subsequent annealing at 650 ^{o}C for 2 h in air (Figure 19). It demonstrated a low turn-on field of about 5.6 V μm^{-1} (here the turn-on field was defined as the value of electric field when an emission current density was 4.5 nA cm^{-2}) according to the FE current density–applied electric field curve (J–E) at a working distance of 60 μm from the anode to the nanorod array serving as the cathode (Figure 33). This FE current–voltage characteristics can be expressed by a simplified Fowler–Nordheim (FN) equation and a field-enhancement factor, β can be defined as $B\phi^{3/2}/\kappa$ according to the FN equation (here ϕ: the work function of cathode material; κ: slope in FN plot).[103] This hncp TiO_2 nanorod array showed a field-enhancement factor β of 8.38×10^{2}. However, a TiO_2 nanorod array with top aggregation for several neighboring nanorods caused by longer deposition displayed a much higher turn-on field of 15.8 V μm^{-1} and lower field-enhancement factor b 3.34×10^{2} (Figure 34). This result indicates that the good FE properties of a periodic TiO_2 hncp nanorod array are mainly attributable to the aligned and periodic hncp nanorod morphology.

When a periodicity of hncp nanorod array increased from 350 nm to 750nm and 1 μm by choosing the colloidal monolayers with different PS sphere sizes during the PLD at the same background gas pressure as before and followed by the same annealing, as presented in Figure 35. The field-enhancement factor decreased with increasing periodicity of the hncp nanorod array (Figure 36a). This is mainly attributed to a decreasing number density of nanorods with an increase in the hncp array periodicity. When the periodicity of the hncp nanorod array was increased from 350 to 750 nm, the turn-on field also increased from 5.6 to 13.0 Vμm^{-1}. When a periodicity further increased to 1 μm, the turn-on field remained about 13.0 Vμm^{-1}. It is evident that the hncp nanorod array with the smallest periodicity of 350nm exhibited the best FE properties in this investigation.

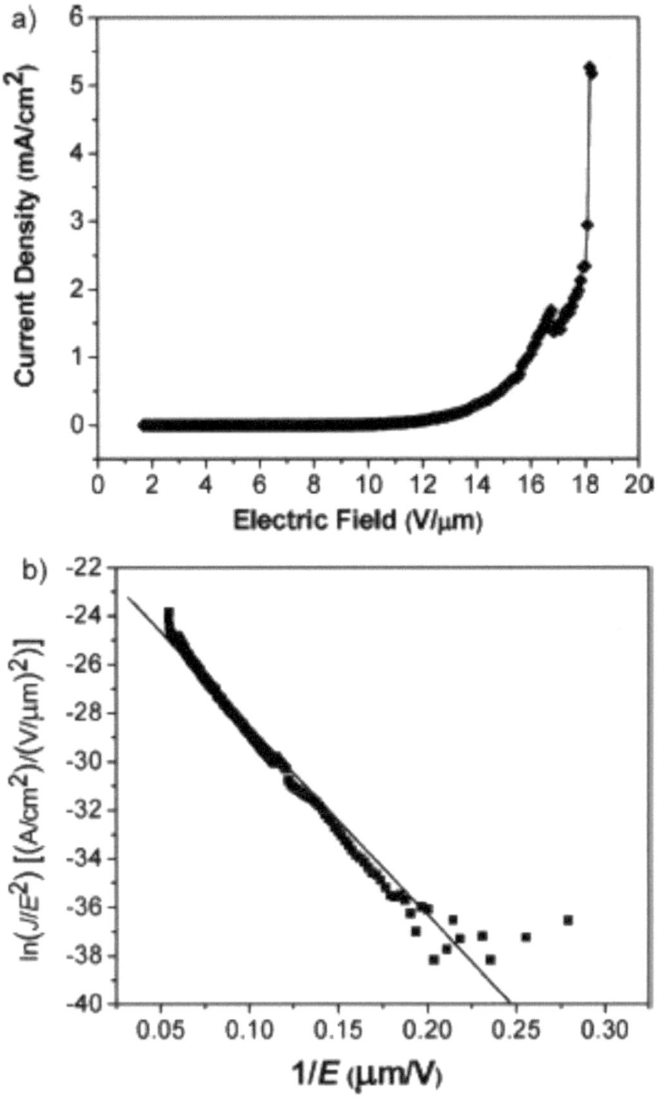

Fig. 33. FE properties of periodic hncp nanorod array by PLD using a colloidal monolayer template with 350nm PS spheres in O_2 at a pressure of 6.7 Pa for 60 min and subsequent annealing in air. a) FE current density–electric field (J–E) curves measured for an hncp TiO_2 nanorod array at an anode–cathode distance of 60 μm. b) Corresponding Fowler–Nordheim (FN) plot.

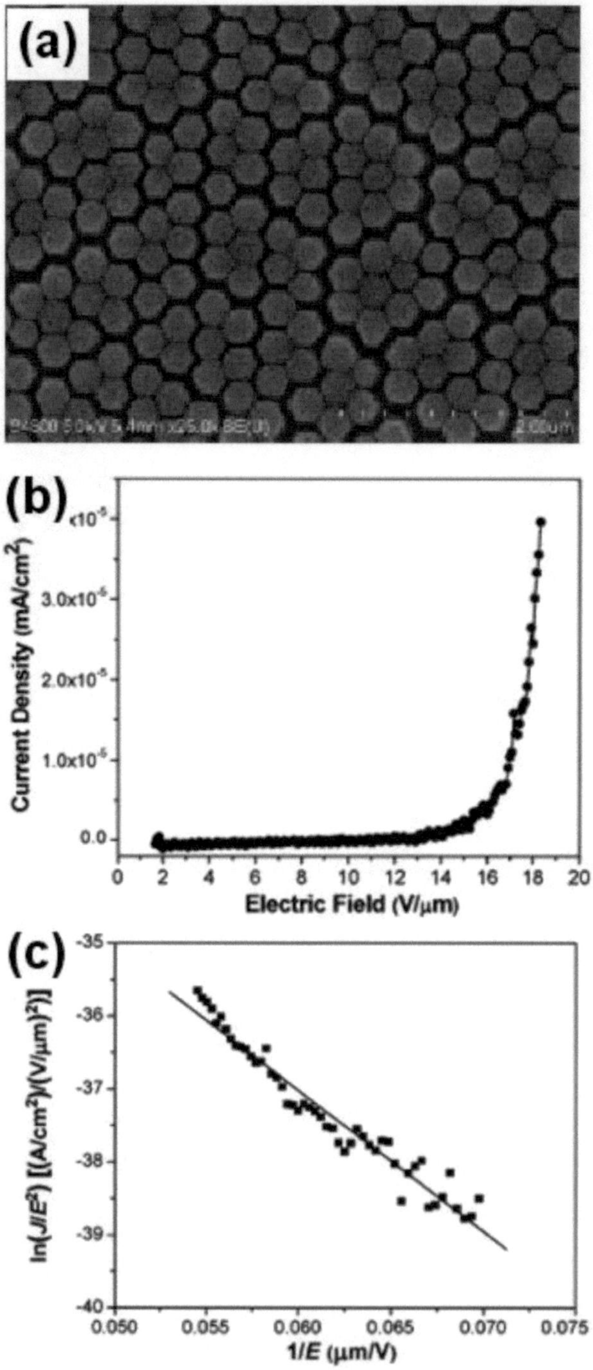

Fig. 34. Top aggregation of a TiO_2 nanorod array obtained PLD assisted colloidal lithography with a longer PLD time (80 min) and subsequent annealing. a) FE-SEM image; b) FE J–E curves measured at an anode–cathode distance of 60 mm for a top-aggregated TiO_2 nanorod array; c) Corresponding FN plot.

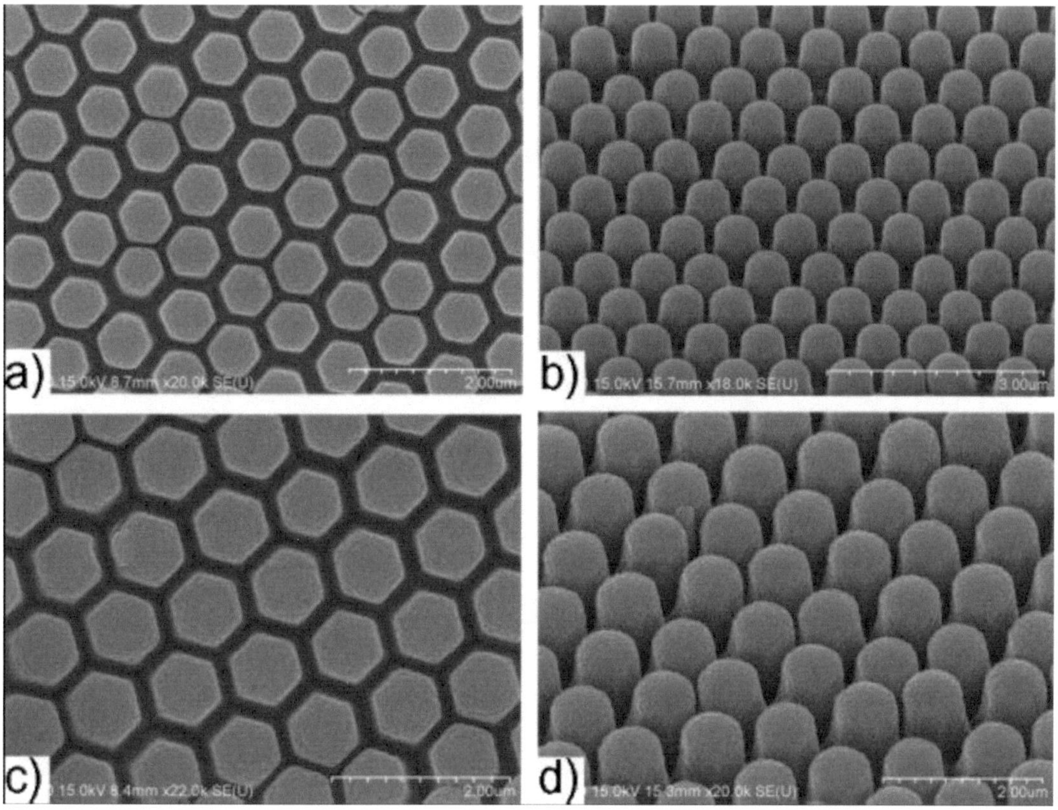

Fig. 35. FE-SEM images of hncp nanorod arrays with different periodicities: 750nm for (a) and (b); 1μm for (c) and (d). PLD was performed in 6.7 Pa of oxygen for 130 min. (a) and (c) are observed from the top, (b) and (d) are observed with a tilt angle of 458.

While a distance between neighboring nanorods can be tuned by changing the background gas pressure during the PLD process if periodicity is fixed to 350 nm for a nanorod array (Figure 22). With increasing this distance of neighboring ones, the field-enhancement factor increased and the turn-on field decreased (Figure 36a). The sample with a small nanorod distance of 20 nm exhibited a relatively low field-enhancement factor of 5.04×10^2 and a high turn-on field of 9.7 V μm^{-1}. When the nanorod distance increased to 50 nm, the FE properties showed enhanced performance with a high field-enhancement factor of 8.38 $\times10^2$ and a low turn-on field of 5.6 V μm^{-1}. When the nanorod distance further increased to 110 nm, the best FE properties with a field-enhancement factor of 9.39 $\times10^2$ and a turn-on field of 5.3 V μm^{-1} were obtained. (Figure 36b) The above results suggest that optimized FE properties can be achieved by increasing the nanorod distance by controlling the experimental parameters. The increased field-enhancement factor b with the increase of nanorod distance can easily be understood as follows if the periodicity of hncp nanorod array is fixed. The field enhancement factor β is generally related to geometry of an emitter and can be expressed as $\beta \propto h/r$, where h is the height and r is the curvature radius of an emitting center. With an increase in the nanorod distance, the effective diameter of an individual nanorod and a curvature radius r would decrease (Figure 37), resulting in an increase of β according to the above relationship.

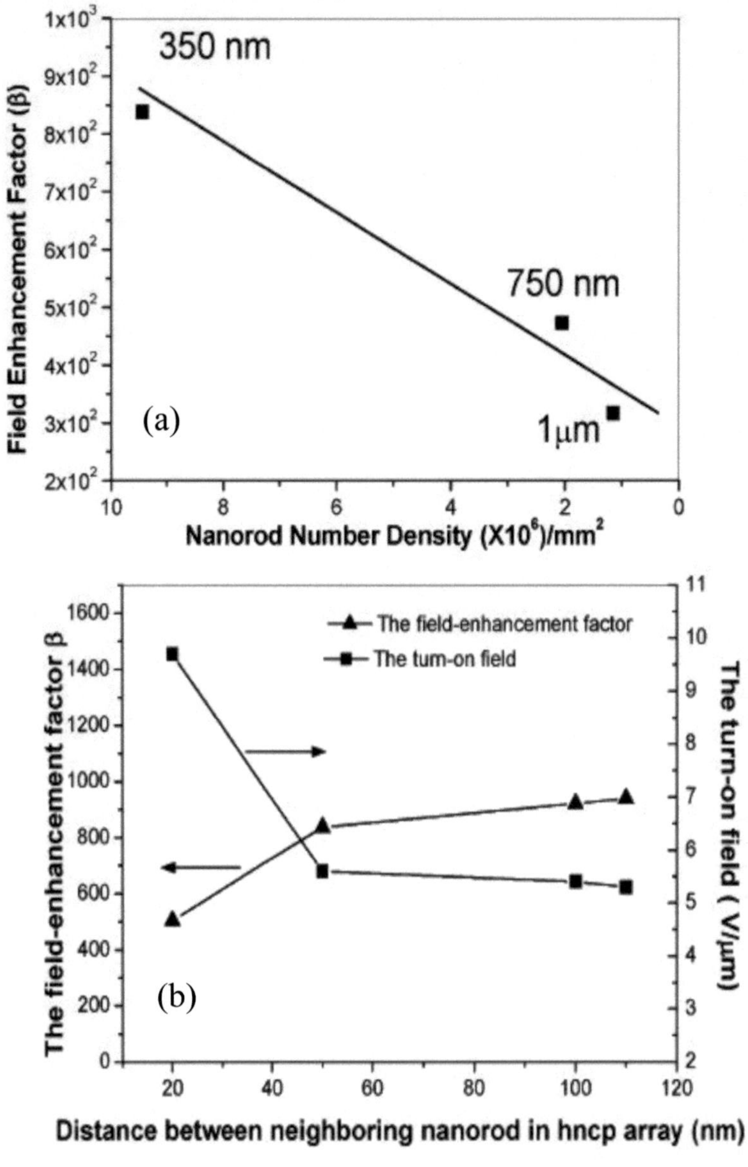

Fig. 36. (a) Field-enhancement factor β changing with increasing periodicity of an hncp nanorod array, (b) Change in field-enhancement factor and turn-on field with varying neighboring nanorod distance in an hncp array.

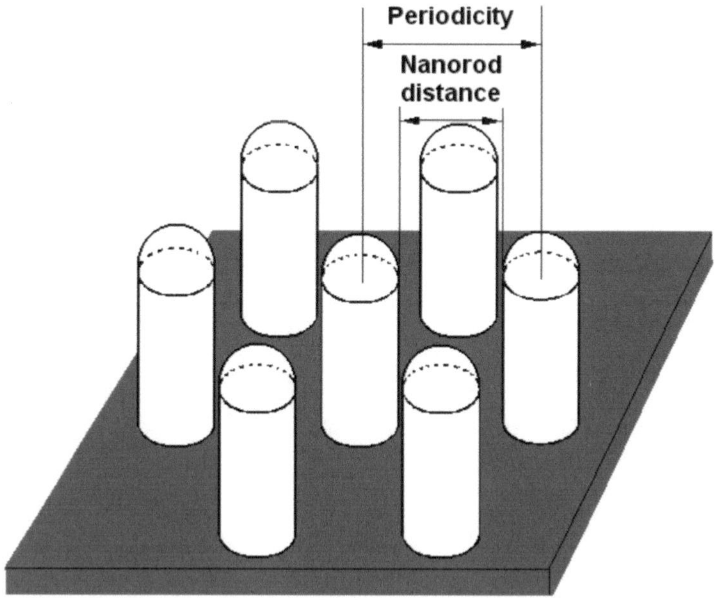

Fig. 37. Schematic illustration of defined nanorod distance in periodic hncp nanorod array.

4.3 Enhanced catalytic properties

The hierarchical micro/nanostructured arrays possess a large specific surface area and hence they might have important application in catalytic fields. For instance, the hcp amorphous TiO_2 micro/nanostructured array on a colloidal monolayer obtained by PLD assisted colloidal lithography demonstrated an enhanced photocatalytic activity (Figure 2). Its photocatalytic performance was estimated based on the decomposition of organic molecules, stearic acid under UV illumination by monitoring the FT-IR spectra.[104,105] The frequencies of 2919 and 2849 cm^{-1} reflect the methylene group asymmetric ($\nu_{asymmCH2}$) and symmetric ($\nu_{symmCH2}$) stretching modes of stearic acid. These values for the methylene group stretching mode are close to those of a crystalline alkane and are typically taken as evidence of the formation of a dense, well-ordered, self-assembled monolayer of stearic acid on the oxide surface.[106-108] Therefore, the photodegradation of stearic acid can be monitor by observing density of these two frequencies. With increasing the UV irradiation time, the vibrational bands of the methylene group gradually decreased and almost completely disappeared after 25 min, as shown in Figure 38a. The decrease in C-H vibrational bands reflects that the stearic acid is gradually photodegraded by such TiO_2 hierarchical micro/nanostructured array films under UV irradiation. Figure 38b shows that degradation curves of a stearic acid film on a silicon wafer, an amorphous TiO_2 film by PLD without using a colloidal monolayer, an hcp amorphous TiO_2 hierarchical micro/nano-rod array on the colloidal monolayer, and an anatase TiO_2 rod array (obtained by annealing hcp amorphous TiO_2 hierarchical micro/nano-rod array on the colloidal monolayer at 650 °C for 2 h. These results indicate that TiO_2 exhibited efficient degradation for stearic acid and that the hcp amorphous TiO_2 hierarchical micro/nano-rod array on a colloidal monolayer demonstrated the best performance compared to the amorphous film and the anatase rod array. Anatase is usually deemed to be more photocatalytically active than the rutile and

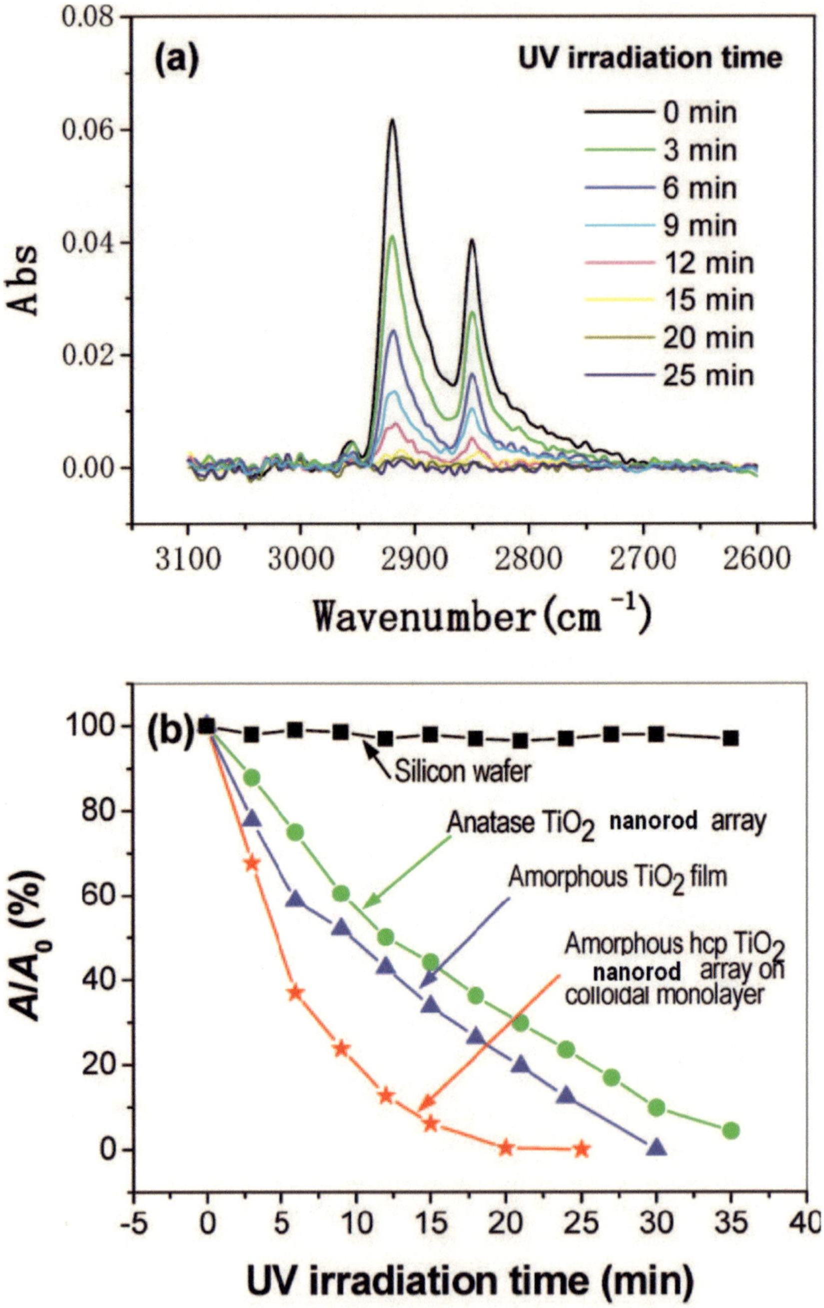

Fig. 38. (a) Photocatalytic activity of an hcp amorphous TiO_2 micro/nano-rod array with a PS colloidal monolayer. (b) Photocatalytic activity evaluation of different substrates based on the absorbance ratio A/A_0 as a function of UV irradiation time. A and A_0 are the absorbance after the UV irradiation and that from the initial surface, respectively.

amorphous TiO_2. However, besides the crystal phase, other factors, including the specific surface area, crystal composition, and material microstructures, also significantly affect the catalytic performance of TiO_2.[109-111] In this case, an amorphous hcp hierarchical micro/nanorod array has porous structures and possesses a much higher specific surface area than that of an anatase rod array, which contributes to better photocatalytic properties. These results suggest that the surface area of TiO_2 is preferable to its crystal structure for enhancing photocatalytic activity. Additionally, a periodic structured array of amorphous TiO_2 can enhance photocatalytic activity better than an amorphous TiO_2 thin film produced by PLD without using a colloidal monolayer. This may be ascribed to special hierarchical structures composed of radiation-shaped nanobranches emanating from a center point on the PS sphere.[112] The combination of superamphiphilicity and photocatalytic activity is helpful in realizing a self-cleaning surface.

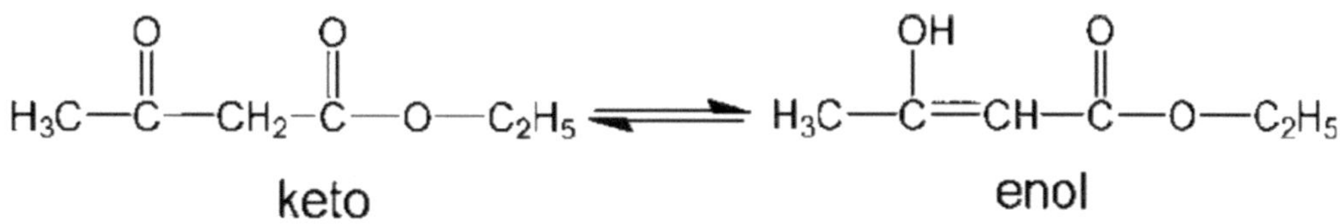

Fig. 39. Isomerization equilibrium of ethyl acetoacetate between the two kinds of isomers, keto and enol.

Additionally, hierarchical alumina mciro/nanostructured arrays demonstrate very excellent catalytic properties for some organic reactions. For example, as we know, ethyl acetoacetate has two kinds of isomers, keto and enol, because of the acidic hydrogen on the active methylene. The corresponding isomerization equilibrium can be depicted in Figure 39. Under common conditions, the keto isomer is more stable than the enol isomer. By the using hierarchical alumina mciro/nanostructured arrays as catalyst, keto isomer might be efficiently converted into the enol isomer. The isomerization can be commonly detected by on-line gas chromatography (GC) or on-line high-performance liquid chromatography (HPLC). Such technique is complicated and, most importantly, not cost-effective. An effective alternative is monitoring the UV-vis spectrum of reaction process. In the keto isomer configuration, there are only two isolated carbonyl groups, with an R absorption band having a quite small ε value in the UV-vis spectrum. For the enol isomer, there is a conjugation system between alkene and carbonyl group, with a K absorption band having a high ε value at around 244 nm in the UV-vis spectrum. This difference makes it convenient to analyze the content and the isomerization process using the UV-vis spectrum. Therefore, the catalytic activity of the hncp alumina was estimated based on the isomerization of ethyl acetoacetate by monitoring the UV-vis absorption spectrum (Figure 40A). The absorption at 244 nm indicates absorption of the conjugation system in the enol isomer as stated above. In the isomerization process, the absorption peak at 244 nm gradually increased and almost completely saturated after 40 min. The increase in absorption at 244 nm indicates that the proportion of enol isomer gradually increased, demonstrating the transition of the keto isomer to the enol isomer and verifying successful isomerization in the presence of hncp alumina. The corresponding control experiments were performed in order to confirm the

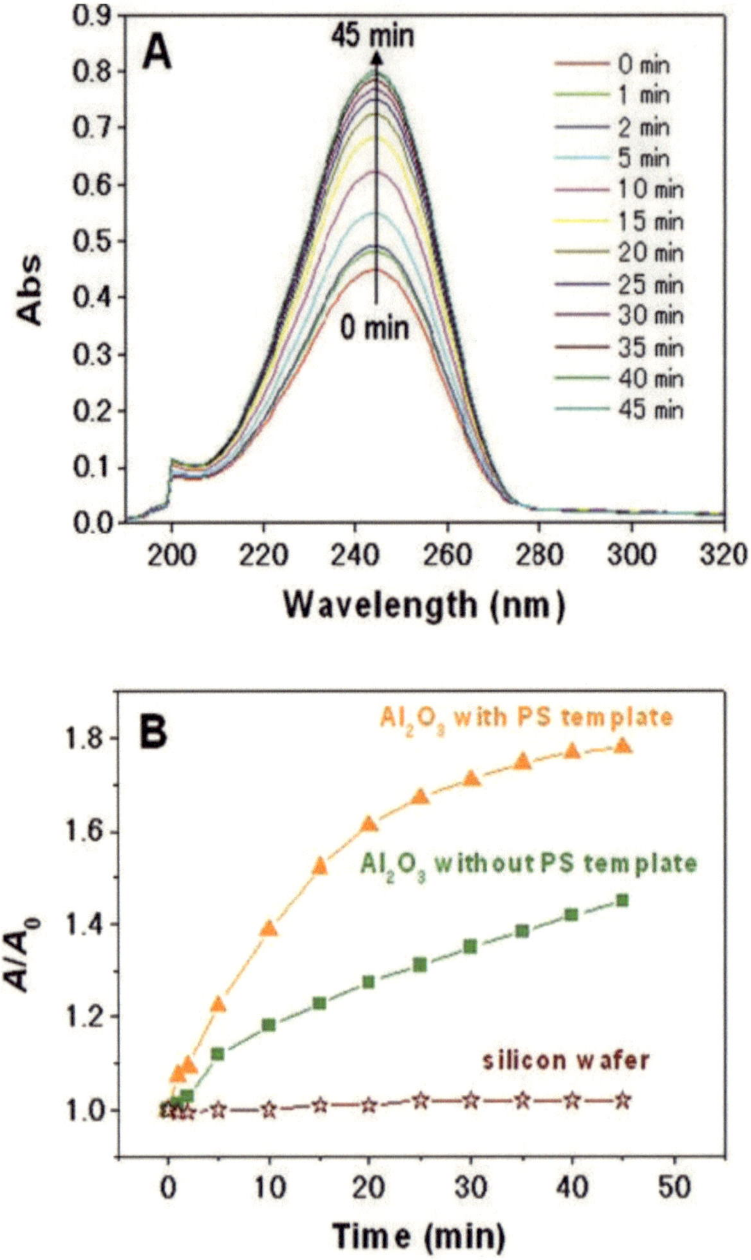

Fig. 40. (A) Catalytic activity of hncp Al_2O_3. (B) Catalytic activity evaluation of different substrates based on the absorbance ratio A/A_0 as a function of reaction time. A and A_0 are the absorbance after a given reaction time and that from the initial solution.

catalytic performance of the hncp alumina for such isomerization. Figure. 40B presents the isomerization process of ethyl acetoacetate in the presence of a silicon wafer and alumina film (by sputtering without using a colloidal monolayer) under the same deposition condition. These results indicate that the hncp alumina exhibited efficient catalytic activity for the isomerization of ethyl acetoacetate. γ-Alumina is generally deemed to be catalytically active. However, besides the crystal phase, other factors, including the specific surface area, crystal composition, material microstructures, and the absence of exotic species from the remnant starting materials, also significantly affect the catalytic performance.[113-116] In this case, the special hierarchical structures of the amorphous hncp alumina array have porous structures and possess a much higher specific surface area than the dense alumina film, which, together with the pure compositions, contributes to better catalytic properties.

5. Conclusions and remarks

The physical deposition assisted colloidal lithography has proven to be a facile, inexpensive, versatile route to construct hierarchical micro/nanostructured arrays with controlled morphologies, sizes, periodicities. The morphologies of these hierarchical micro/nanostructured arrays can be tuned by controlling the experimental conditions, including deposition time, background gas pressure in the vacuum chamber, periodicity of colloidal monolayer template etc. Compared with chemical routes, the physical deposition are more suitable for preparing high quality micro/nanostructured arrays with uniform morphologies. These special structures possess morphology- or size-dependent properties, such as superamphiphilicity, superhydrophobicity, photocatalytic activity, field emission etc., which have important applications in devices, microfluidic devices, field emitters, solar cells etc. Compared to development of fabrication strategies of micro/nanostructured arrays, investigation of morphology- or parameter- properties and micro/nanodevices dependent on them is not so much. The more micro/nanodevices based on these structures will be hoped, and it might be realized under researchers` efforts in the future.

6. Acknowledgements

This work was financially supported by the Natural Science Foundation of China (Grant Nos. 50831005, 10974203), provincial Natural Science Foundation of Anhui (Grant No. 11040606M62), and the National Basic Research Program of China (973 Program, Grant No. 2011CB302103).

7. References

[1] M. Morariu, N. Voicu, E. Schäffer, Z. Lin, T. P. Russell, U. Steiner, *Nat. Mater,.* 2003, 2, 48.
[2] C. H. Ye, L. D. Zhang, X. S. Fang, Y. H. Wang, P. Yan, J. W. Zhao, *Adv. Mater.,* 2004, 16, 1019.
[3] S. V.Dorozhkin, *J. Mater. Sci.: Mater. Med.,* 2007, 18, 363.
[4] G. Duan, W. Cai,Y. Luo, Y. Li, Y. Lei, *Appl. Phys. Lett.,* 2006, 89, 181918.
[5] J. Y. Lao, J. G. Wen, D. Z. Wang, Z. F. Ren, *Nano. Lett.,* 2002, 2, 1287.
[6] P. X. Gao, Y. Ding, Z. L. Wang, *Nano. Lett.,* 2003, 3, 1315.
[7] R. Yan, D. Gargas, P. Yang, *Nat. Photonics,* 2009, 3, 569.
[8] E. Roduner, *Chem. Soc. Rev.,* 2006, 35, 583.
[9] C. M. Cobley, J. Chen, E. C. Cho, L. V. Wang, Y. Xia, *Chem. Soc. Rev.,* 2011, 40, 44.
[10] Z. L., Wang, *Adv. Func Mater.,* 2008, 18, 3553.
[11] R. A. Varin, L. Zbroniec, M. Polanski, J. Bystrzycki, *Energies,* 2011, 4, 1.
[12] Smith, H. I.; Schattenburg, M. L. *IBM J. Res. DeV.* 1993, *37,* 319.

[13] Stroscio, J. A.; Eigler, D. M. *Science* 1991, *254*, 1319.
[14] Liu, G.-Y.; Xu, S.; Qian, Y. *Acc. Chem. Res.* 2000, *33*, 457.
[15] Piner, R. D.; Zhu, J.; Xu, F.; Hong, S.; Mirkin, C. A. *Science* 1999, *283*, 661.
[16] Xia, Y. N.; Whitesides, G. M. *Langmuir* 1997, *13*, 2059.
[17] Kumar, A.; Whitesides, G. M. *Appl. Phys. Lett.* 1993, *63*, 2002.
[18] Quist, A. P.; Pavlovic, E.; Oscarsson, S. *Anal. Bioanal. Chem.* 2005, *381*,591.
[19] Kim, E.; Xia, Y.; Whitesides, G. M. *Nature* 1995, *376*, 581.
[20] L. Gao, T. J. McCarthy, *Langmuir*, 2006, *22*, 2966.
[21] Micheletto, R.; Fukuda, H.; Ohtsu, M. *Langmuir* 1995, *11*, 3333.
[22]Rakers, S.; Chi, L. F.; Fuchs, H. *Langmuir* 1997, *13*, 7121.
[23] Denkov, N. D.; Velev, O. D.; Kralchevsky, P. A.; Ivanov, I. B.; Yoshimura, H.; Nagayama, K. *Langmuir* 1992, *8*, 3183.
[24] Denkov, N. D.; Velev, O. D.; Kralchevsky, P. A.; Ivanov, I. B.; Yoshimura, H.; Nagayama, K. *Nature* 1993, *361*, 1303.
[25] Hulteen, J. C.; Van Duyne, R. P. *J. Vac. Sci. Technol.* 1995, *13*, 1553.
[26] Ozin, G. A.; Yang, S. M. *Adv. Funct. Mater.* 2001, *11*, 95.
[27] Jiang, P.; McFarland, M. J. *J. Am. Chem. Soc.* 2004, *126*, 13778.
[28] Wang, D.; Möhwald, H. *Adv. Mater.* 2004, *16*, 244.
[29] Mihi, A.; Ocana, M.; Mí guez, H. *Adv. Mater.* 2006, *18*, 2244.
[30] Li, Y.; Cai, W. P.; Duan, G. T.; Sun, F. Q.; Cao, B. Q.; Lu, F. *Mater.Lett.* 2005, *59*, 276.
[31] Antony, S.; Dimitrov, A. S.; Nagayama, K. *Langmuir* 1996, *12*, 1303.
[32] Jiang, P.; Bertone, J. F.; Hwang, K. S.; Colvin, V. L. *Chem. Mater.*1999, *11*, 2132.
[33] Im, S. H.; Kim, M. H.; Park, O. O. *Chem. Mater.* 2003, *15*, 1797.
[34] Kitaev, V.; Ozin, G. A. *Adv. Mater.* 2003, *15*, 75.
[35] Choi, W. M.; Park, O. O. *Nanotechnology* 2006, *17*, 325.
[36] G. Zhang, D. Wang, *Chem. Asian J.*, 2008, 4, 236.
[37] Y. Li, N. Koshizaki, W. Cai, *Coord. Chem. Rev.*, 2010, 255, 357.
[38] Y. Li, W. Cai, G. Duan, *Chem. Mater.*, 2008, 20, 615.
[39] L. Li, T. Zhai, H. Zeng, X. Fang, Y. Bando, D. Golberg. *J. Mater. Chem.*, 2011, 21, 40.
[40] J. Zhang, B. Yang, *Adv. Funct. Mater.*, 2010, 20, 3411.
[41] S.-M. Yang, S.G. Jang, D.-G. Choi, S. Kim, H.K. Yu, *Small*, 2006, 2, 458.
[42] Sun, F.; Cai, W.; Li, Y.; Duan, G.; Nichols, W. T.; Liang, C.; Koshizaki, N.; Fang, Q.; Boyd, I. W. *Appl. Phys. B: Lasers Opt.* 2005, *81*, 765.
[43] Burmeister, F.; Schäfle, C.; Matthes, T.; Böhmisch, M.; Boneberg, J.; Leiderer, P. *Langmuir* 1997, *13*, 2983.
[44] Pacifica, J.; Gómez, D.; Mulvaney, P. *AdV. Mater.* 2005, *17*, 415.
[45] Jensen, T. J.; Duval, M. L.; Kelly, K. L.; Lazarides, A. A.; Schatz, G. C.; Van Duyne, R. P. *J. Phys. Chem. B* 1999, *103*, 9846.
[46] Malinsky, M. D.; Kelly, K. L.; Schatz, G. C.; Van Duyne, R. P. *J. Phys. Chem. B* 2001, *105*, 2343.
[47] Hulteen, J. C.; Treichel, D. A.; Smith, M. T.; Duval, M. L.; Jensen, T. J.; Van Duyne, R. P. *J. Phys. Chem. B* 1999, *103*, 3854.
[48] Tan, B. J. Y.; Sow, C. H.; Koh, T. S.; Chin, K. C.; Wee, A. T. S.; Ong, C. K. *J. Phys. Chem. B* 2005, *109*, 11100.
[49] Sort, J.; Glaczynska, H.; Ebels, U.; Dieny, B.; Miersig, M.; Rybczynski, J. *J. Appl. Phys.* 2004, *95*, 7516.

[50] Sun, F. Q.; Cai, W. P.; Li, Y.; Cao, B. Q.; Lei, Y.; Zhang, L. D. *Adv. Funct. Mater.* 2004, *14*, 283.
[51] Sun, F. Q.; Yu, J. C.; Wang, X. C. *Chem. Mater.* 2006, *18*, 3774.
[52] Li, Y.; Cai, W. P.; Duan, G. T.; Sun, F. Q.; Lu, F. *Appl. Phys. A: Mater. Sci. Process.* 2005, *81*, 269.
[53] Li, Y.; Cai, W. P.; Cao, B. Q.; Duan, G. T.; Li, C. C.; Sun, F. Q.; Zeng, H. B. *J. Mater. Chem.* 2006, *16*, 609.
[54] Sun, F. Q.; Cai, W. P.; Li, Y.; Cao, B. Q.; Lei, Y.; Zhang, L. D. *Mater. Sci. Technol.* 2005, *21*, 500.
[55] Sun, F. Q.; Cai, W. P.; Li, Y.; Jia, L. C.; Lu, F. Adv. *Mater.* 2005, *17*, 2872.
[56] Sun, F. Q.; Cai, W. P.; Li, Y.; Cao, B. Q.; Lu, F.; Duan, G. T.; Zhang, L. D. Adv. *Mater.* 2004, *16*, 1116.
[57] Cao, B. Q.; Cai, W. P.; Sun, F. Q.; Li, Y.; Lei, Y.; Zhang, L. D. *Chem. Commun.* 2004, 1604.
[58] Cao, B. Q.; Sun, F. Q.; Cai, W. P. *Electrochem. Solid-State Lett.* 2005, *8*, G237.
[59] Duan, G. T.; Cai, W. P.; Li, Y.; Li, Z. G.; Cao, B. Q.; Luo, Y. Y. *J. Phys.Chem. B* 2006, *110*, 7184.
[60] Duan, G. T.; Cai, W. P.; Luo, Y. Y.; Sun, F. Q. *Adv. Funct. Mater.* 2007, *17*, 644
[61] Wang, X.; Lao, C.; Graugnard, E.; Summers, C. J.; Wang, Z. L. *Nano Lett.* 2005, *5*, 1784.
[62] Yan, F.; Goedel, W. A. *Nano Lett.* 2004, *4*, 1193.
[63] Duan, G. T.; Cai, W. P.; Luo, Y. Y.; Li, Z. G.; Lei, Y. *J. Phys. Chem.B* 2006, *110*, 15729.
[64] Li, Y.; Cai, W. P.; Duan, G. T.; Cao, B. Q.; Sun, F. Q. *J. Mater. Res.* 2005, *20*, 338.
[65] Venkatesh, S.; Jiang, P. *Langmuir* 2007, *23*, 8231.
[66] Li, Y.; Huang, X. J.; Heo, S. H.; Li, C. C.; Choi, Y. K.; Cai, W. P.; Cho, S. O. *Langmuir* 2007, *23*, 2169.
[67] Huang, X. J.; Li, Y.; Im, H. S.; Yarimaga, O.; Kim, H. J.; Jang, D. Y.; Cho, S. O.; Cai, W. P.; Choi, K. Y. *Nanotechnology* 2006, *17*, 2988.
[68] Correa-Duarte, M. A.; Kosiorek, A.; Kandulski, W.; Giersig, M. *Small* 2006, *2*, 220.
[69] Li, Y.; Lee, E. J.; Cho, S. O. *J. Phys. Chem. C* 2007, *111*, 14813.
[70] Li, Y.; Li, C. C.; Cho, S. O.; Duan, G. T.; Cai, W. P. *Langmuir* 2007, *23*, 9802.
[71] Duan, G. T.; Cai, W. P.; Luo, Y. Y.; Li, Y.; Lei, Y. *Appl. Phys. Lett.* 2006, *89*, 181918.
[72] Qi, L. M.
[73] Li, Y.; Sasaki, T.; Shimizu, Y.; Koshizaki, N. *J. Am. Chem. Soc.* 2008, *130*, 14755.
[74] Li, Y.; Koshizaki, N.; Shimizu, Y.; Li, L.; Gao, S. Sasaki, T. *ACS Appl. Mater. & Interfaces.* 2009, *1*, 2580.
[75] Nam, H. J.; Sasaki, T.; Koshizaki, N. *J. Phys. Chem. B* 2006, *110*, 23081.
[76] Brett, M. J.; Hawkeye, M. M. *Science* 2008, *319*, 1192.
[77] Robbie, K.; Beydaghyan, G.; Brown, T.; Dean, C.; Adams, J.; Buzea, C. *Rev. Sci. Instrum.* 2004, *75*, 1089.
[78] Zhao, Y.-P.; Ye, D.-X.; Wang, G.-C.; Lu, T.-M. *Nano Lett.* 2002, *2*, 351.
[79] Hrudey, P. C. P.; Szeto, B.; Brett, M. J. *Appl. Phys. Lett.* 2006, *88*, 251106.
[80] Kesapragada, S. V.; Victor, P.; Nalamasu, O.; Gall, D. *Nano Lett.* 2006, *6*, 854.
[81] He, Y. P.; Fu, J. X.; Zhang, Y.; Zhao, Y. P.; Zhang, L. J.; Xia, A. L.; Cai, J. W. *Small* 2007, *3*, 153.
[82] Zhou, C. M.; Gall, D. *Appl. Phys. Lett.* 2006, *88*, 203117.
[83] Robbie, K.; Sit, J. C.; Brett, M. J. *J. Vac. Sci. Technol., B: Microelectron. Nanometer Struct.sProcess., Meas., Phenom.* 1998, *16*, 1115.

[84] Vick, D.; Tsui, Y. Y.; Brett, M. J.; Fedosejevs, R. *Thin Solid Films* 1999, *350*, 49.
[85] Li, Y.; Sasaki, T.; Shimizu, Y.; Koshizaki, N. Small, 2008, 4, 2286.
[86] Li, Y.; Fang, X.; Koshizaki, N.; Sasaki, T.; Li, L.; Gao, S.; Shimizu, Y.; Sasaki, T.; Bando, Y.; Golberg, M. Adv. Funct. Mater. 2008, 19, 2467.
[87] Li, L.; Koshizaki, N. J. Mater. Chem., 2010, 20, 2972
[88] Li, L.; Li, Y.; Gao, S.; Koshizaki, N. J. Mater. Chem., 2009, 19, 8366.
[89] Gao, S.; Koshizaki, N.; Li, Y.; Li, L. J. Mater. Chem., 2011, 21, 2087.
[90] A. Nakajima, K. Hashimoto, T. Watanabe, Monatsh. Chem. 132 (2001) 31.
[91] T. Sun, L. Feng, X. Gao, L. Jiang, Acc. Chem. Res. 38 (2005) 644.
[92] X.-M. Li, D. Reinhoudt, M. Crego-Calama, Chem. Soc. Rev. 36 (2007) 1350.
[93] X.-J. Huang, J.-H. Lee, J.-W. Lee, J.-B. Yoon, Y.-K. Choi, Small 4 (2008) 211.
[94] X.-J. Huang, D.-H. Kim, M. Im, J.-H. Lee, J.-B. Yoon, Y.-K. Choi, Small, 2008, 5, 90
[95] R.N. Wenzel, J. Phys. Colloid Chem. 1949, 53, 1446.
[96] R. Wang, K. Hashimoto, A. Fujishima, M. Chikuni, E. Kojima, A. Kitamura, M. Shimohigoshi, T. Watanabe, Adv. Mater. 1998, 10, 135.
[97] W.-L. Min, B. Jiang, P. Jiang, Adv. Mater. 2008, 20, 3914.
[98] Y. Li, J. Zhang, S. Zhu, H. Dong, Z. Wang, Z. Sun, J. Guo, B. Yang, J. Mater. Chem. 2009, 19, 1806.
[99] Y.-R. Lin, H.-P. Wang, C.-A. Lin, J.-H. Hea, J. Appl. Phys. 2009, 106, 114310.
[100] A. Cassie and S. Baxter, Trans. Faraday Soc., 1944, 40, 546.
[101] S.S. Fan, M.G. Chapline, N.R. Franklin, T.W. Tombler, A.M. Cassell, H.J. Dai, Science 283 (1999) 512.
[102] X.S. Fang, Y. Bando, U.K. Gautam, C.H. Ye, D. Golberg, J. Mater. Chem. 18 (2008) 509.
[103] R.H. Fowler, L.W. Nordheim, Proc. R. Soc. London, Ser. A 119 (1928) 173.
[104] Wang, K. X.; Yao, B. D.; Morris, M. A.; Holmes, J. D. *Chem. Mater.* 2005, *17*, 4825.
[105] Zhang, X. T.; Jin, M.; Liu, Z. Y.; Tryk, D. A.; Nishimoto, S.; Murakami, T.; Fujishima, A. *J. Phys. Chem. C* 2007, *111*, 14521.
[106] Gao, W.; Dickinson, L.; Grozinger, C.; Morin, F. G.; Reven, L. *Langmuir* 1996, *12*, 6429.
[107] Nuzzo, R. G.; Dubois, L. H.; Allara, D. L. *J. Am. Chem. Soc.* 1990, *112*, 558.
[108] Gawalt, E. S.; Avaltroni, M. J.; Koch, N.; Schwartz, J. *Langmuir* 2001, *17*, 5736.
[109] Fujishima, A.; Zhang, X. T. *C. R. Chimie* 2006, *9*, 750.
[110] Macak, J. M.; Zlamal, M.; Krysa, J.; Schmuki, P. *Small* 2007, *3*, 300.
[111] Wu, C. Z.; Lei, L. Y.; Zhu, X.; Yang, J. L.; Xie, Y. *Small* 2007, *3*, 1518.
[112] Li, H. X.; Bian, Z. F.; Zhu, J.; Zhang, D. Q.; Li, G. S.; Huo, Y. N.; Li, H.; Lu, Y. F. *J. Am. Chem. Soc.* 2007, *129*, 8406.
[113] J. Li, R. Yan, B. Xiao, D. T. Liang and L. Du, Environ. Sci. Technol., 2008, 42, 6224;
[114] L. B. Sun, J. Yang, J. H. Kou, F. N. Gu, Y. Chun, Y. Wang, J. H. Zhu and Z. G. Zou, Angew. Chem., Int. Ed., 2008, 47, 3418;
[115] C. Misra, American Chemical Society Monograph, in Industrial Alumina Chemicals, American Chemical Society, Washington, DC,1986, vol. 184;
[116] B. R. Baker and R. M. Pearson, J. Catal., 1974, 33, 265.

Part 4

Periodic Structures for Diffractive Optics and Surface Wetting

6

Design of Circular Dammann Grating: Fabrication and Analysis

Fung Jacky Wen and Po Sheun Chung
Department of Electronic Engineering, City University of Hong Kong
China

1. Introduction

The term "diffraction" has been defined by Sommerfeld as follows (E. Hecht, 2002): any deviation of light rays from rectilinear paths which cannot be interpreted as reflection or refraction. Diffraction is caused by the confinement of the lateral extent of a wave and is most importantly when that confinement is comparable in size to the wavelength of the radiation being used. The first accurate report and description of such a phenomenon was made by Grimaldi and was published in the year 1665. Later, in 1678, Christian Huygens expressed the intuitive conviction that of each point on the wavefront of a disturbance were considered to be a new source of a secondary spherical disturbance. This technique, however, ignores most of each secondary wavelet and retains only that portion common to the envelope. As a result of this inadequacy, this principle is unable to account for the details of the diffraction process. The difficulty was resolved by Fresnel and Kirchhoff with his addition of the concept of interference in late 18th century. These types of diffraction are known for two centuries in the form of diffraction gratings which periodically modulate the incident wave-front. An ideal grating generates a set of waves, called diffraction orders that propagate into discrete directions. The diffraction angles θ_m are given by the well-known grating equation $\sin\theta_m = m\lambda/D$, where λ is the wavelength of light and D is the grating period, is shown in Fig. 1. The amplitudes of the diffraction orders are determined by the structure of the periodic modulation. The demand for electromagnetic analysis is arising together with the advance of the fabrication technology. A significant step was taken as well since the development of computers. It can compute such a complicated wave field numerical analysis from which the design a desired grating is much easier.
The DOE is designed for splitting the input beam into M diffraction orders. For array illuminators, equal power intensity with high uniformity is necessary. To achieve this special feature, a periodic nature together with binary phase structure, which was proposed by Dammann in the early 70's, is one of the solutions (H. Dammann & K. Gortler, 1971; H. Dammann & E. Klotz, 1977). To analyze the performance of the Dammann grating, we employ the Burckhardt, Kaspar and Knop (BKK) method and TE-polarized dependent mode is normally assumed (C. B. Burckhardt, 1966). In order to have an easier understanding, the entire Dammann grating has the identities in terms of periodic, symmetric and binary structure. The total normalized diffraction efficiency is just the sum of all required

normalized diffraction power. Normally, for reducing the zero-order intensity and polarization dependent loss, the ratio of period/wavelength should be as large as possible. We have also employed Dammann Grating as a beam splitter (J. F. Wen & P. S. Chung, 2007; J. F. Wen & P. S. Chung, 2007).

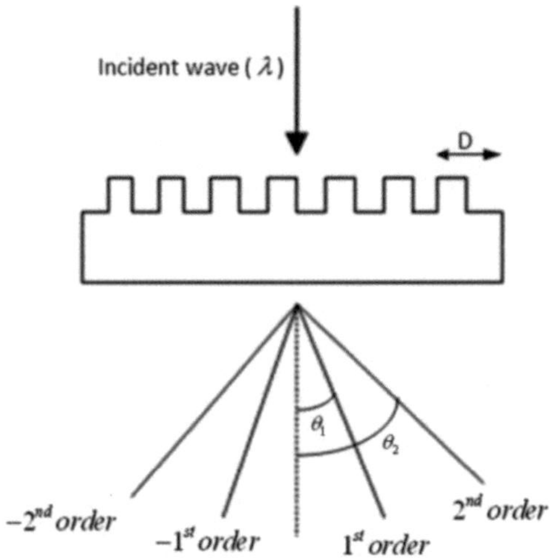

Fig. 1. Diffraction Angle for DOE.

2. Introduction to Circular Dammann Grating (CDG)

In previous section, we know that light could be diffracted into multiple orders with equal intensities and high uniformity with the theory of Dammann grating. The diffracted spots could be one-dimensional or two dimensional depending on the applications. For some applications, optical systems need circular images, e.g. laser free space communication system (J. Jia, C. Zhou & L. Liu, 2003), fast focal length measurement (S. Zhao, J. F. Wen & P. S. Chung, 2007; S. Zhao & P. S. Chung, 2007) and usage in DFB laser (C. Wu *et al*, 1991; T. Erdogan *et al,* 1992). We have then further extended the Dammann grating into CDG. The CDG is one of the possible candidates which can produce circular beams in ring-shape at the image plane.

Zhou et al (C. Zhou, J. Jia & L. Liu, 2003) first proposed the concept of CDG in 2003 based on the modulation of the Bessel function using a binary phase annulus mask. The phase and radius of each annulus can be modified so that the intensity at the far field can be manipulated. However the CDG does not have the periodic nature as most of the gratings required and therefore, it is only a DOE and equal separation cannot also be achieved. Recently Zhao and Chung (S. Zhao & P. S. Chung, 2007) proposed a new design method for the periodic CDG using the coefficients of the circular sine series for generating equal-intensity and equal-spacing of optical rings, which means those infinite circular periods can be repeated. We have also presented another two novel approaches based on the concept of circular spot rotation (J. F. Wen, S. Y. Law & P. S. Chung, 2007), Hankel transform (J. F. Wen & P. S. Chung, 2008) to achieve the same objectives as mentioned above with higher efficiency and uniformity. In our research, we have employed the Circular Dammann Grating into angle, area and distance measurements respectively (J. F. Wen & P. S. Chung, 2008; J. F. Wen, Z. Y. Chen & P. S. Chung, 2008; J. F. Wen, Z. Y. Chen, & P. S. Chung, 2010).

2.1 Circular spot rotation method

The concept of generating CDG using this method is based on the theory of conventional Dammann Grating (C. Zhou & L. Liu, 1995). The cross section and first order spectrum of CDG is shown in Fig. 2 and 3 respectively. We assumed that if the diffraction spots rotate 360 degrees continuously, circular rings will be formed. Fig. 4 and 5 illustrate these ideas. Table 1 shows some numerical results. For the formula of Circular Dammann Grating, the intensities among different orders will then be as follows:

$$n^{th}order: \; P_n = |M(n)|^2 = \frac{1}{n^2\pi^2}\{[\sum_{k=1}^{N}(-1)^k \cos(2\pi n r_k)]^2 + [\sum_{k=1}^{N}(-1)^k \sin(2\pi n r_k)]^2\} \tag{2.1}$$

$$0^{th} order: \quad P_0 = \left[2\sum_{k=1}^{N}(-1)^k r_k + 1\right]^2 \tag{2.2}$$

The overall normalized efficiency is therefore given as,

$$\eta = \sum_{n=-\infty}^{\infty} nP_{2n-1} \tag{2.3}$$

The feature size, which is the minimum distance between two different continuous transition points, is

$$\Delta = \min|r_{k+1} - r_k| \tag{2.4}$$

The uniformity is defined as,

$$uni = \frac{\max(I_n) - \min(I_n)}{\max(I_n) + \min(I_n)} \tag{2.5}$$

The radius (R) of each ring with the focal length (f) of converging lens will then be

$$R_n = \frac{n\lambda}{D} f \tag{2.6}$$

The numerical solutions with near optimum efficiency and uniformity of CDG are listed in Table 1.

Circle number	Normalized Transition Points in Half of Period*				Efficiency	Uniformity	Feature Size
1	0.5				0.81	0	0.5
2	0.20525	0.29067	0.5		0.72	0.00006	0.08542
3	0.11649	0.24024	0.26741	0.38396	0.74	0.0002	0.02717
	0.5						
4	0.099104	0.18382	0.26295	0.32925	0.8	0.00014	0.00508
	0.4196	0.49492	0.5				

*The other half period could be calculated using $x_{k+i} = x_i + 0.5$

Table 1. Some Numerical Solutions of CDG by Spot Rotation Method.

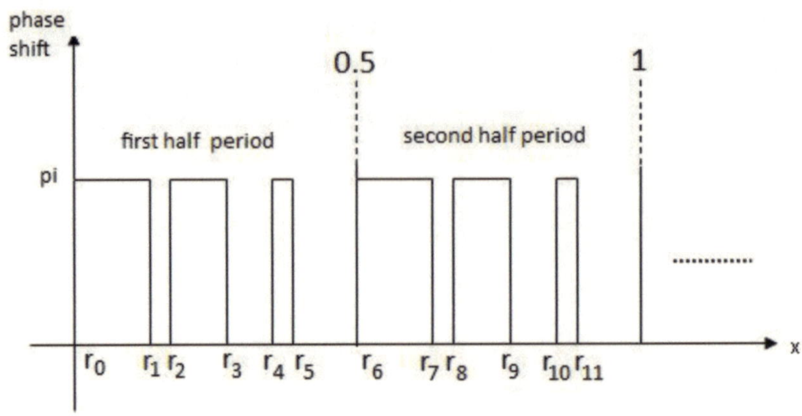

Fig. 2. Cross Section of CDG.

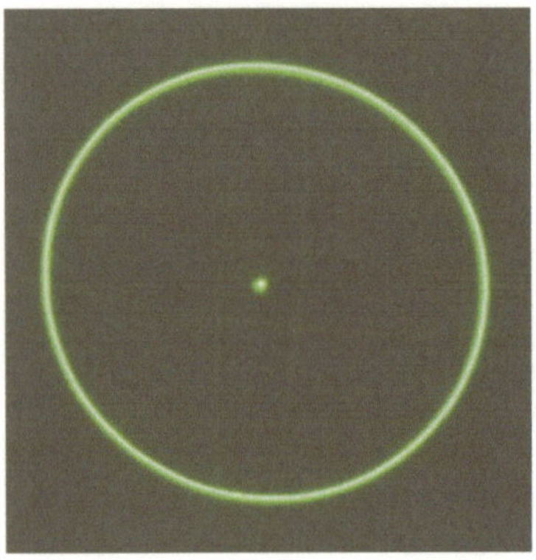

Fig. 3. First order CDG spectrum.

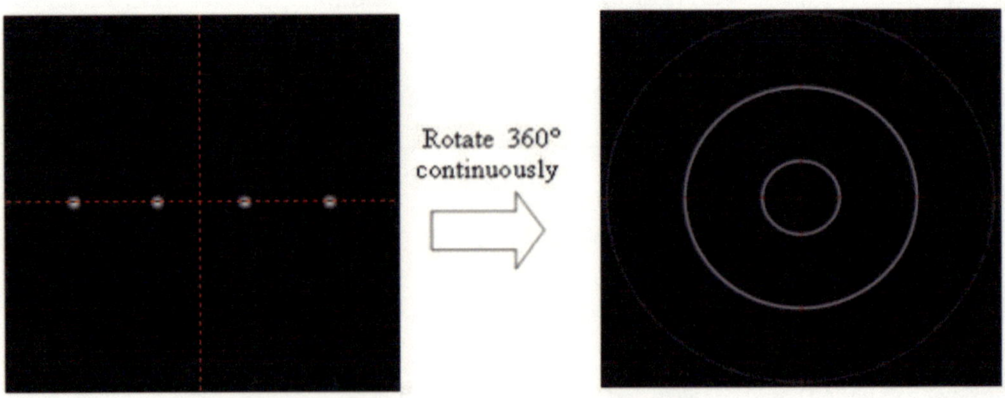

Fig. 4. Concept of Circular Spot Rotation Method's type CDG.

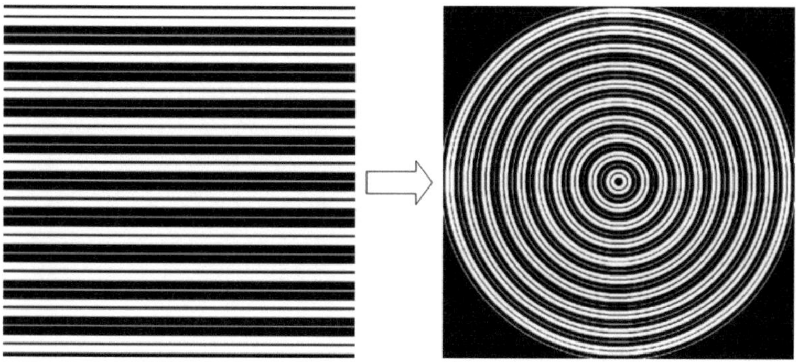

Fig. 5. Circular Spot Rotation Method's type CDG Profile Design.

2.2 Hankel transform

The concept of this grating is based on the theory of Hankel transform. It is a two dimensional Fourier transform with a radially symmetric integral and is also called the Fourier-Bessel transform (F. Bowman, 1958). The cross section of CDG is still same as Fig. 2. The problem of side lobe effect is one of the major concerns in both Zhao and our circular rotation methods. The effect is shown in Fig. 6. In this section, we present another novel approach based on the Bessel function together with Hankel transform in symmetry structure to achieve the same objectives with better performance in terms of efficiency and uniformity while the lower side lobe power could be obtained. Table 2 shows some numerical results. The comparisons are shown in Fig. 7-9.

The intensities equations for each diffraction order are given as follows:

$$n^{th}order: \; I_q = |M(q)|^2 = \frac{1}{q^2}\left|2\sum_{k=1}^{N}(-1)^{k+1} r_k J_1(2\pi q r_k) + r_N J_1(2\pi q r_N)\right|^2 \tag{2.7}$$

$$0^{th}order: \; I_0 = \left[2\sum_{k=1}^{N}(-1)^k r_k + 1\right]^2 \tag{2.8}$$

Circle Number	Normalized Transition Points in Half of Period*			Feature[1]	Efficiency[2] (Main Lobe)	Overall Efficiency[3]	Uniformity
1	0.39763	0.4907	0.5	0.0093	0.665	0.931	
2	0.091023	0.3266	0.5	0.0610	0.612	0.855	0.043
3	0.10036	0.23569	0.26741	0.0353	0.60	0.866	0.05
	0.38396	0.5					
4	0.099572	0.1838	0.26314	0.0038	0.69	0.99	0.036
	0.32894	0.41929	0.49616				
	0.5						

Note:
1: Feature is the minimum distance between two subsequent transition points
2: It describes the sum of powers in terms of main lobes only.
3: It describes the sum of all powers including the main and side lobes.
*: The other half period could be calculated using $x_{k+i} = x_i + 0.5$

Table 2. Some Numerical Solutions of CDG by Hankel Transform.

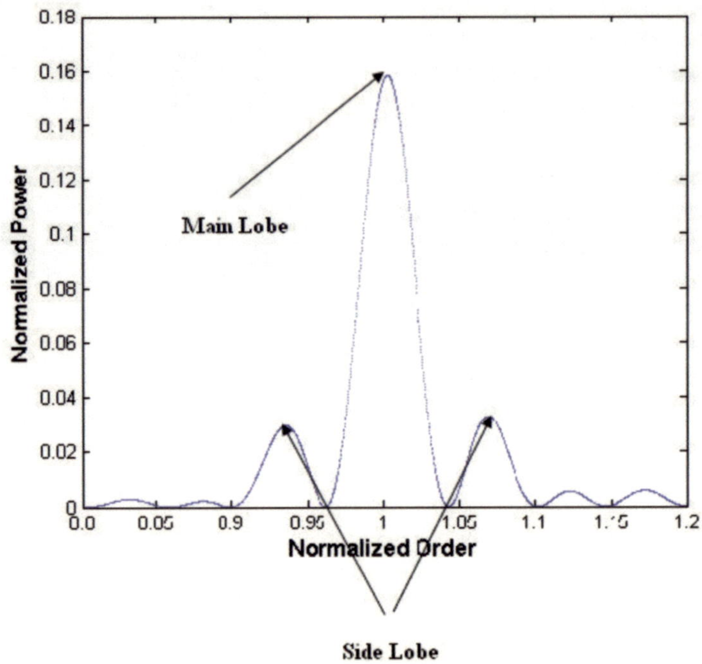

Fig. 6. Side Lobe Effect of CDG.

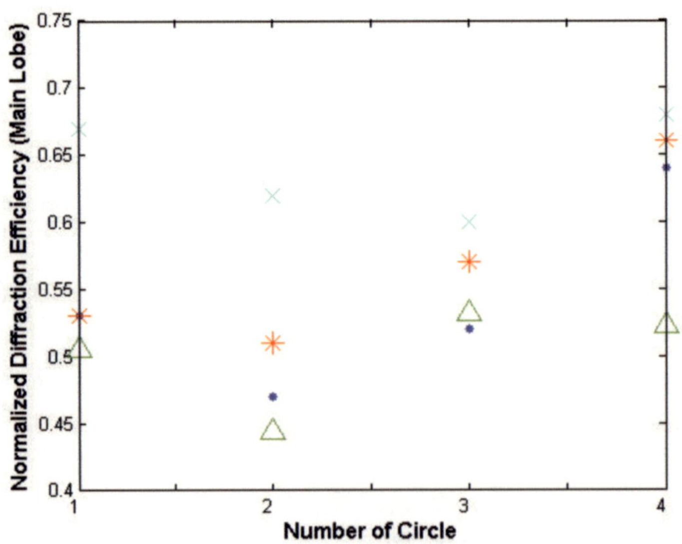

Fig. 7. Main Lobe Intensities Comparisons among Four Different Methods (Triangle: Zhao's symmetric method, Dot: Zhao's asymmetric method, Star: Spot Rotation Method, Cross: Hankel Transform).

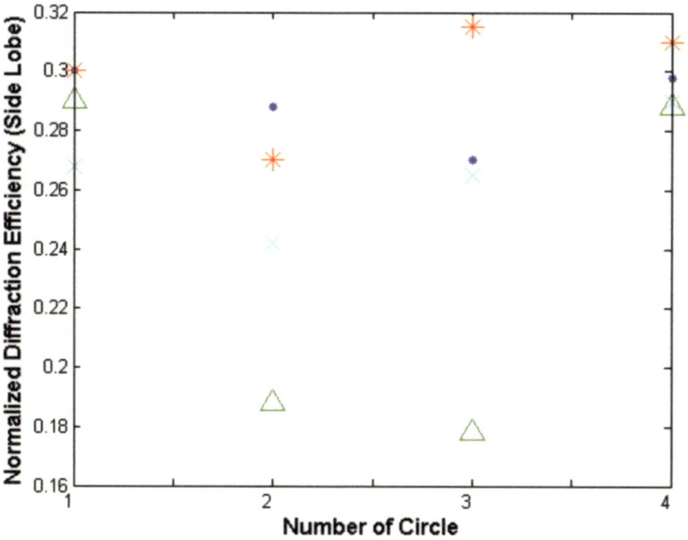

Fig. 8. Side Lobe Intensities Comparisons among Three Different Methods (Triangle: Zhao's symmetric method, Dot: Zhao's asymmetric method, Star: Spot Rotation Method, Cross: Hankel Transform).

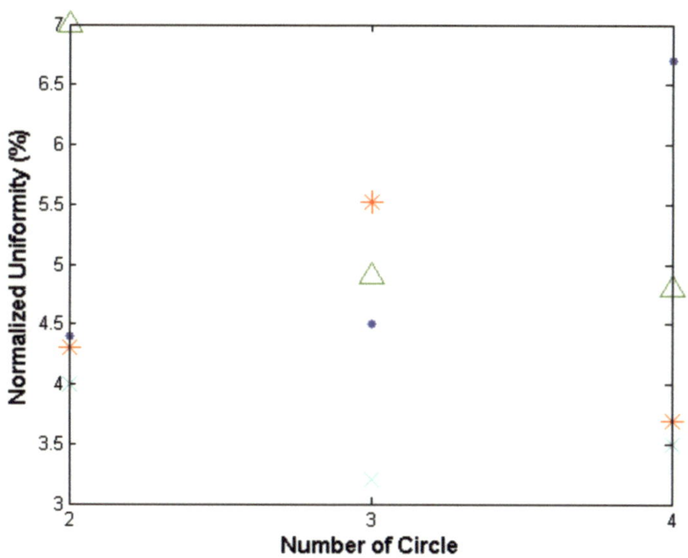

Fig. 9. Uniformities Comparisons among Three Different Methods (Triangle: Zhao's symmetric method, Dot: Zhao's asymmetric method, Star: Spot Rotation Method, Cross: Hankel Transform).

The overall normalized efficiency, feature size, uniformity and radius are defined same as equations 2.3-2.6.
Side lobe is existed around the main lobe in every diffracted order, is shown in Fig. 6, no matter which method is applied. It is because our design is not able to fully express the concept of circular phase modulation, as the circular profile can not be completely decomposed into a square pixel representation and also finite number of periods happened in practical case. Thus side lobe existed. The lobe separation is not only governed by number of period (*ND*), but also controlled by the input wavelength and the focal length of the lens. With the assistance of diffraction theory, the final equation is then defined, i.e.

$$s = \frac{2\lambda}{3ND} f \tag{2.9}$$

Fig. 10-12 shows this relationship. From these figures, we can conclude that 100 periods is the optimum solution.

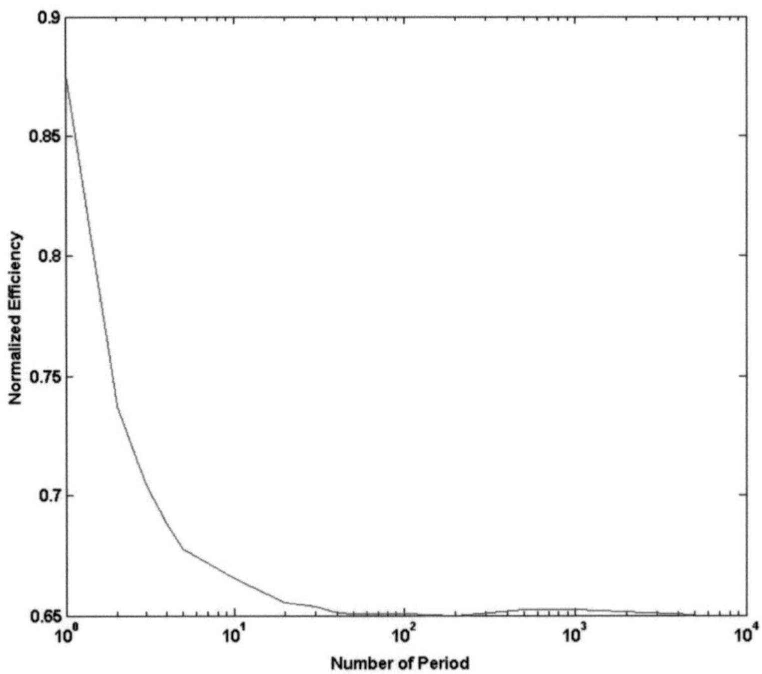

Fig. 10. Number of Period against efficiency.

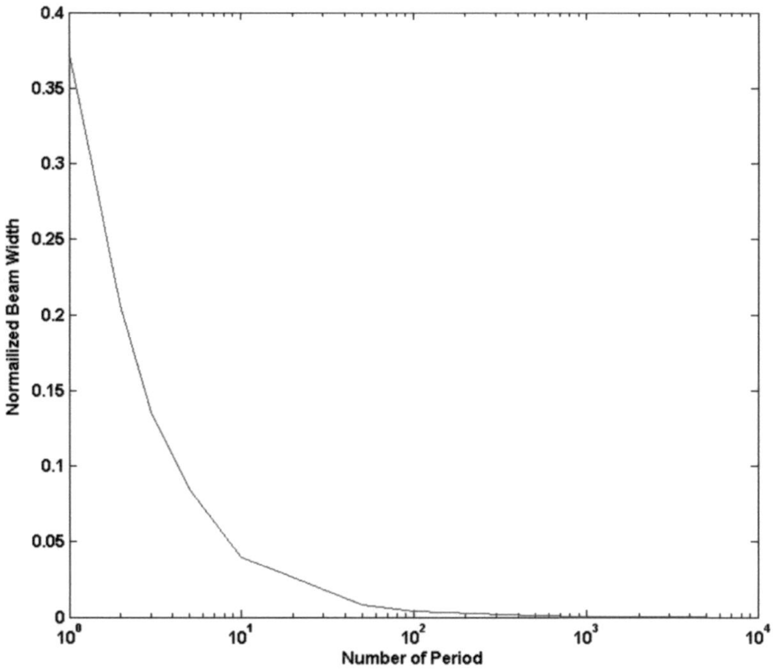

Fig. 11. Number of period against beam width (FWHM).

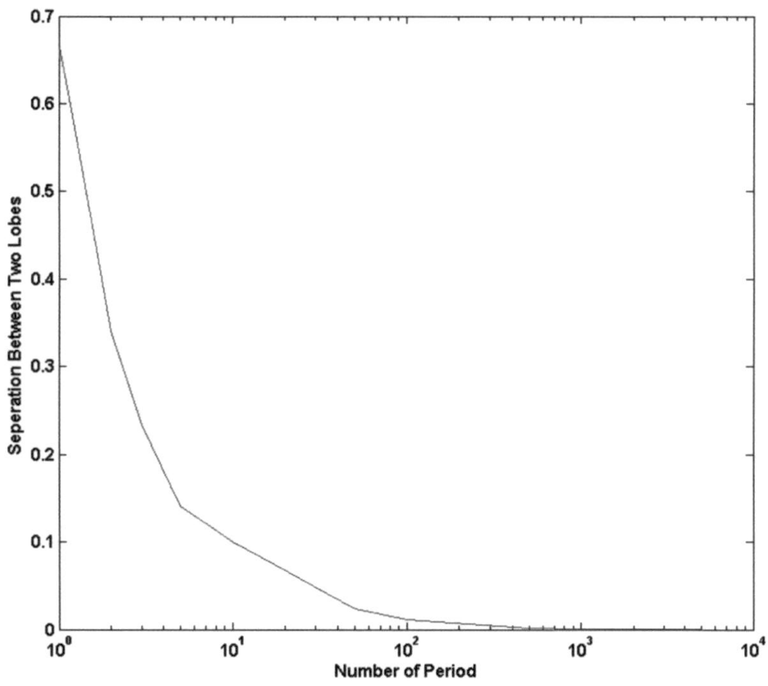

Fig. 12. Number of Period against separation between two lobes.

3. Fabrication

Many different fabrication methods exist for diffraction grating. Most of these techniques can be grouped into two main categories: lithographic techniques and electron beam writing. Lithographic techniques (J. Turunen, A Vasara, J Westerholm, G Jin and A Salin, 1990) use light sensitive polymers at the top of the substrate along with controlled etching or deposition methods. For standard e-beam lithography (Masato Okano, Hisao Kikuta, Yoshihiko Hirai, Kazuya Yamamoto, and Tsutom Yotsuya, 2004), an e-beam exposure contains enough of an electron dose that the exposed regions of e-beam resist are fully cleared during the development process. This can be used to produce different thicknesses of e-beam resist simply by varying the dose. The e-beam approach can generate the finest features, a serval tens of nm. However, because of the small size of the electron beam, it is extremely time consuming to expose a pattern of large size of sample.

3.1 Chemical etching fabrication

In previous sections, we have understood the fundamental theories of diffraction and a number of design approaches for CDG. To fabricate these gratings, we have to maintain the desired grating shape which is critical to guarantee high transmission of the order of interest. An optimal depth is very important since existing zero order transmission will happen afterwards. Other defeats, such as mask misalignment, sharpness of the profile, will scatter light to higher orders. Because of the size of the features and the need for flexibility of fabrications, lithographic method is an optimal for many types of DOE fabrication [M. T. Gale, 1997; J. M. Miller et al, 1993; C. Zheng, 2005). A simplified version of the photolithographic processing sequence is shown in Fig. 13.

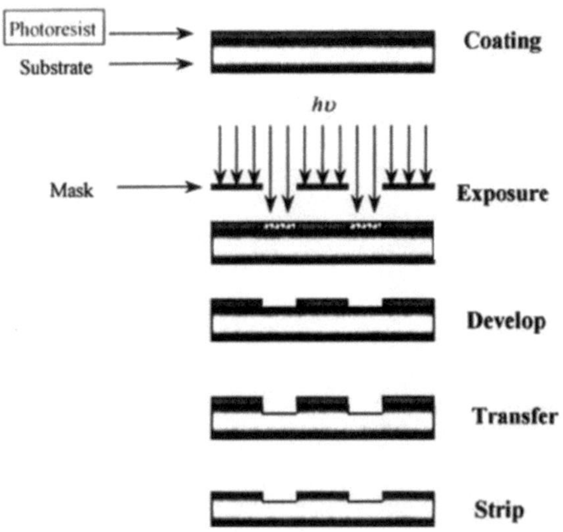

Fig. 13. Flow Diagram of chemical etching.

Once we confirm the pattern which is determined from the design process, the depth of the diffractive phase structure, which is determined by the wavelength of the incident light and the refractive indices of the substrate and surrounding material, can be calculated. The depth can be calculated by

$$h = \frac{\lambda}{2(n_1 - n_0)} \tag{3.1}$$

where λ is the input wavelength and n_1, n_0 are the indices of refraction of the substrate material and the surrounding medium at the operating wavelength respectively. The substrate in using is quartz and the refractive index is shown in Fig. 14.

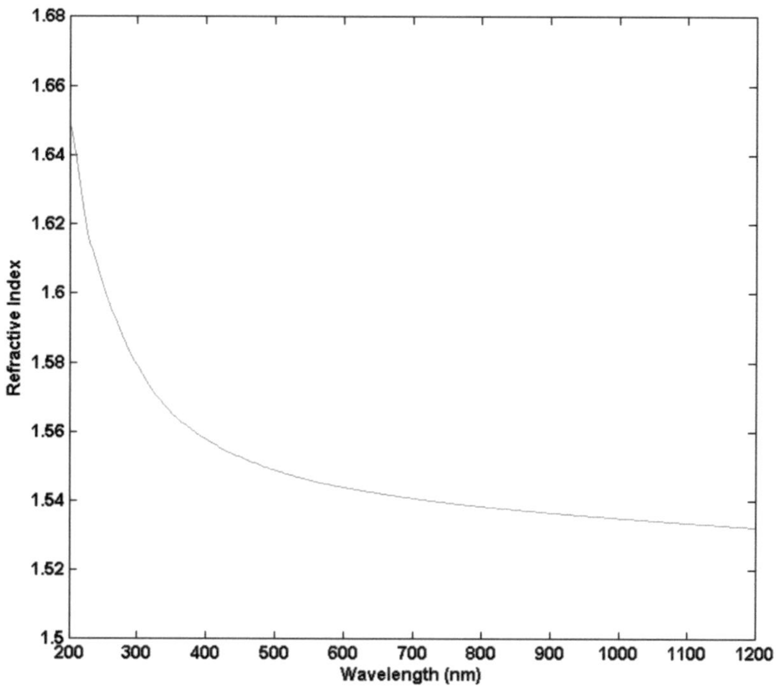

Fig. 14. Refractive Index of quartz.

Most lithographic masks are binary transmission masks. That is, they contain alternating clear and opaque areas. The opaque areas mean the Chromium remains on top of the substrate. These masks are usually made by forming the pattern in a light-sensitive photoresist on top of thin chrome later on the glass mask. Once the photoresist is developed, the chrome, where the photoresist has been removed, was protected. The mask pattern is exposed using optical pattern generators with controllable beam size. A variety of file formats, e.g. GDSII, CIF and BMP, can be used. The machine in our lab is "Microtech LW405". The positioning accuracy is 1um and the minimum linewidth is 0.8um. These patterns are transformed into pixel forms with a given dimension and the magnified mask pattern is shown in Fig. 15. The accuracy of patterning curve is related to the wavefront error introduced by the required shape approximation. Increasing the number of pixels can help to improve the accuracy. However, as we expect, the higher degree of accuracy can result of more time spending and the amount of data is also increased. Once the mask is fabricated, which is shown in Fig. 16, we can move on to the next step.

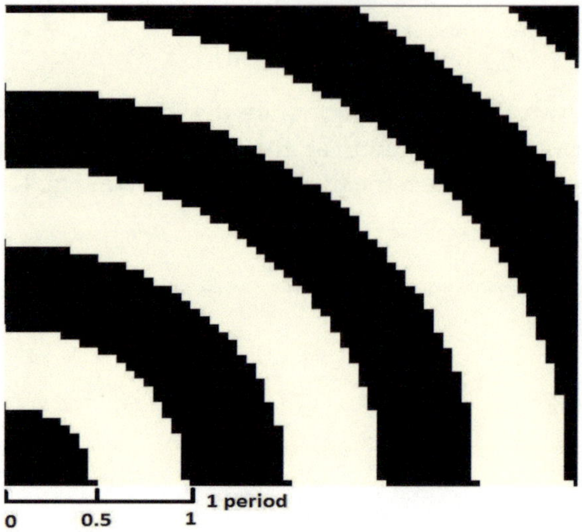

Fig. 15. Magnified CDG pattern.

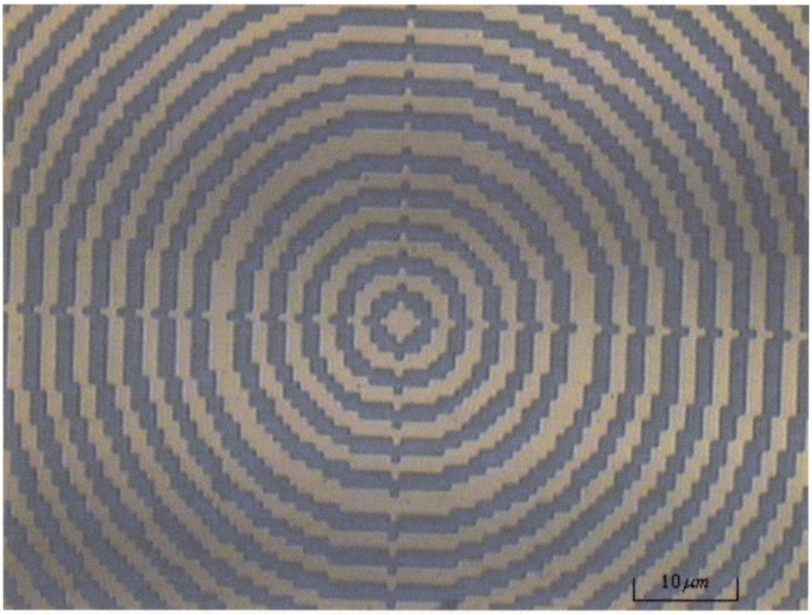

Fig. 16. CDG Pattern with the number of periods 100.

Photolithographic methods are based on the use of photoresist create relief structures on substrate surfaces. This structure is used to protect the underlying substrate during subsequent processing steps. Photoresists, which is light sensitive polymer, can be either positive, where the exposed resist dissolves, or they can be negative, where the exposed resist remains after development. Positive photoresists (e.g. SPR6112B) is used in the following processing example. After substrate is cleaned, the first step is to coat the substrate with a thin (typically microns) layer of photoresist. It can be generated by spin coat the wafer. During spin coating, liquid photoresist is distributed uniformly around the wafer as it rotates at high rates. The spin curve of this photoresist is shown in Fig. 17.

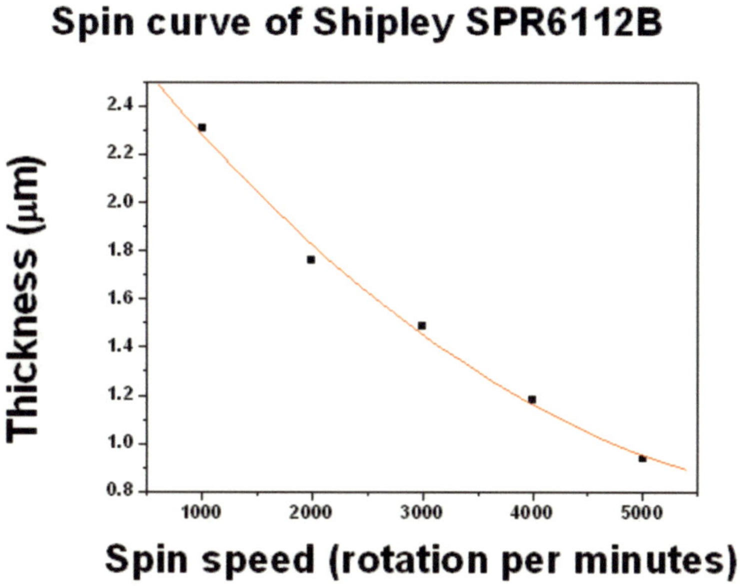

Fig. 17. Spin Curve for SPR6112B.

In order to have a better evaporation of phtoresist, it is often to be heated at a hundred degree Celsius. The final thickness of the photoresist layer is controlled by a combination of the viscosity of the photoresist and the spin speeds used during the coating process. The next step is the exposure of the photoresist. Patterns will be formed at the photoresist layer using Aligner, a uniform ultraviolet light source. The mask can be placed in contact with the photorsist layer for a high resolution, 1:1 transfer of the image scale, this process is referred to as contact printing. The exposed photoresist is washed away after exposure. Developing solution and developing time also affect the fidelity of the resulting lithographic image. Over, or underdevelopment will decrease the fidelity of the image pattern. For example, while exposed resist dissolves much faster than unexposed areas, the un-irradiated areas will also begin to lose photoresist if the development time is too long. The simplified illustration is shown in Fig. 18.

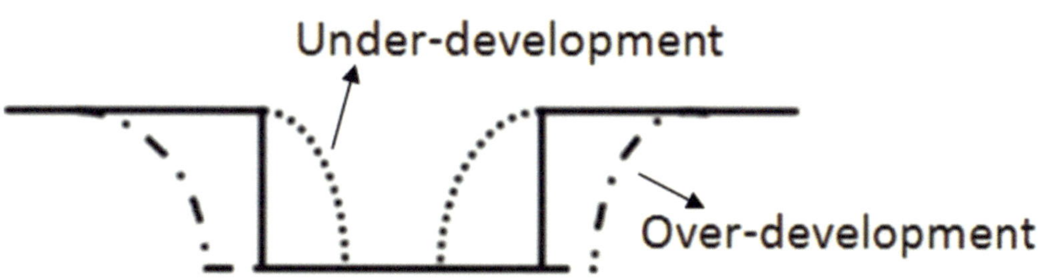

Fig. 18. Diagram of different types of development.

The wafer has been normally spin coated with Chromium with SiO_2 as the substrate. This material can be functioned from the ultraviolet through the near infrared due to its own transmission properties and low coefficient of thermal expansion. Normally, a photoresist pattern serves as a wall that protects the area under the photoresist during etching. In our experience, a single photoresist layer without any other protection cannot stand for a long period of time. In this case, we add one more chromium layer which the purpose is same as photoresist. As a result, only the areas not covered by the chromium are removed during the etching process.

The wet etching technique is isotropic, which its definition is the etching rate is equal in all directions. This desires for our grating case, particularly for applications requires sharply defined and vertical features. Standard diffractive etch chemistries for silica, usually HF acid, can act as an etching solution. In our experiment, we use another solution, named as Fluorosilicic acid. This etching solution offers a straightforward way to create smooth features in glass materials. In addition fluorosilicic acid etching is considerably safer than conventional hydrofluoric acid (HF) etching. Normally, the temperature controlled heating stage was set at 60°C, which gave an etching solution temperature of 30°C (±0.5°C). Fluorosilicic acid, 100 ml (20% concentration) was allowed to obtain an etch rate 56 nm/min in our experience.

3.2 Electron beam lithography

Electron beam lithography (EBL) is based on the principle that some polymers are sensitive to electrons and can be patterned by electron exposure, which is very much like the other lithography. PMMA (Poly Methyl Methacrylate) is normally used in EBL. Its own low exposure time and low resolution are the limitations for our case. We therefore shifted to another resist named "ZEP 520A" by ZEON corporation. ZEP520A is high performance positive EB resists which show high resolution and dry etch resistance. The spin curve and refractive index of this resist are shown in Fig. 19 and 20. The schematic of electron beam system is shown in Fig. 21.

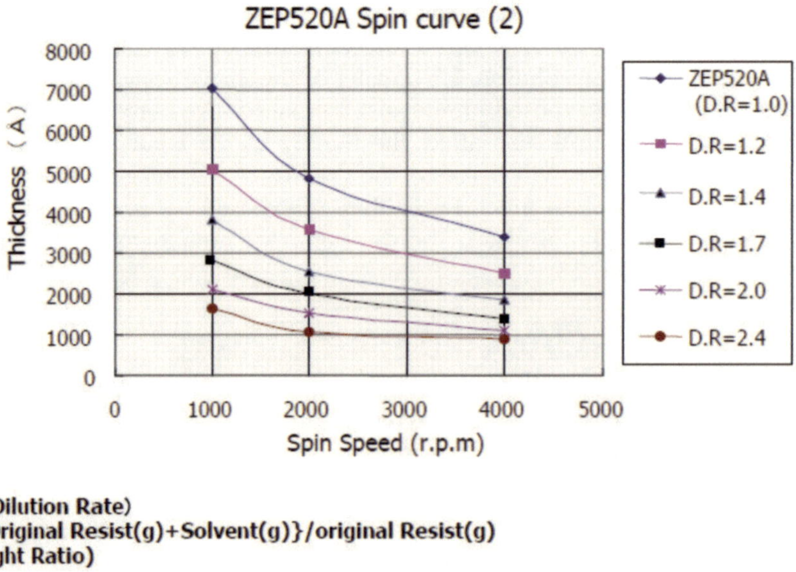

Fig. 19. Spin Curve for ZEP520A with Dilution Rate (provided by ZEON).

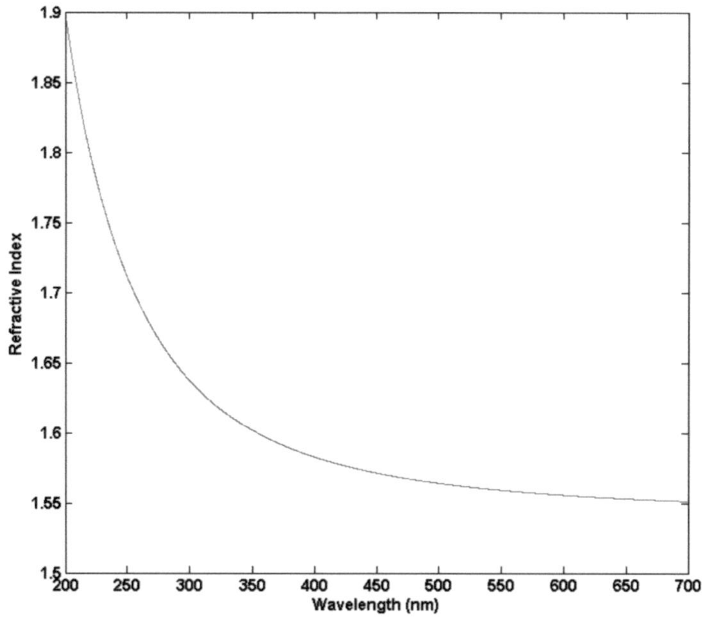

Fig. 20. Refractive Index of ZEP520A (provided by ZEON).

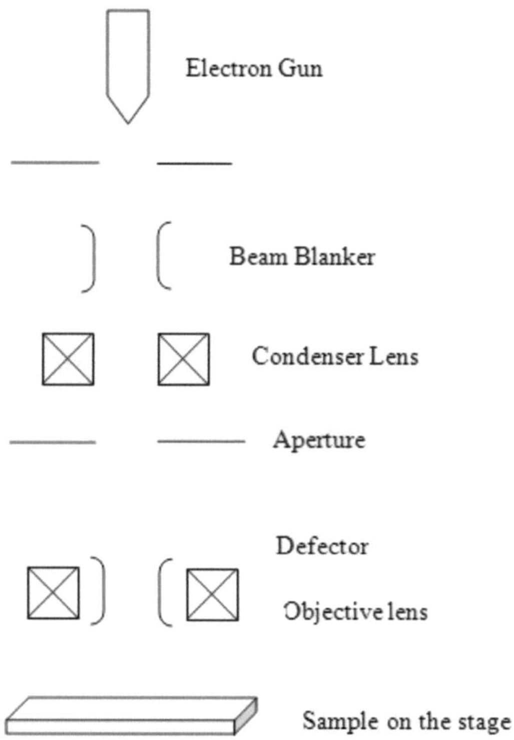

Fig. 21. Simplified System of EBL.

At the top, it consists of an electron gun, a condenser lens to allow changes in the current and corresponding beam diameter, an objective lens to focus the beam on the wafer and a deflector to scan the e-beam around within the field. The sample is placed below on a motorized stage so that it can be patterned by the desired profile. More detailed descriptions of the different types of sources, lenses and the various other components can be found in many textbooks (C. Zheng, 2005). The simplified flow diagram is shown in Fig. 22. The CDG sample is also shown in Fig. 23. The machine in our lab is "Crestec CABL-9510C".

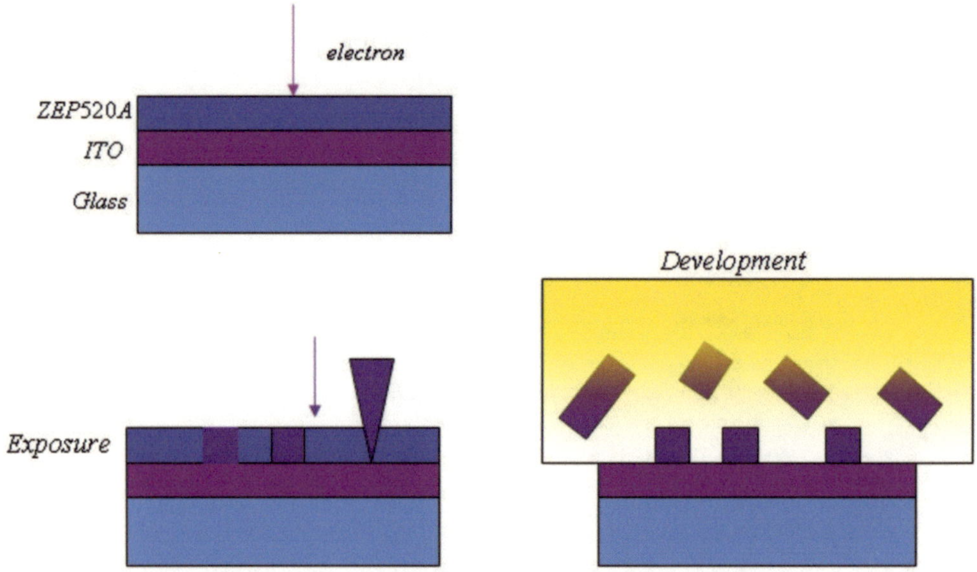

Fig. 22. Flow diagram of EBL.

Fig. 23. CDG sample using EBL.

4. Conclusion

CDG is affected by the different errors, such as etched depth errors, feature errors, grating period deviations and non vertical side wall angle. Most of these errors will affect the distribution of light into diffracted order. The effects of variation in phase depth and grating duty cycle for a grating are shown in Fig. 24-26 respectively. In general, from the following figures, we can conclude that for getting over 60% efficiency, 100nm variation is allowed in both techniques.

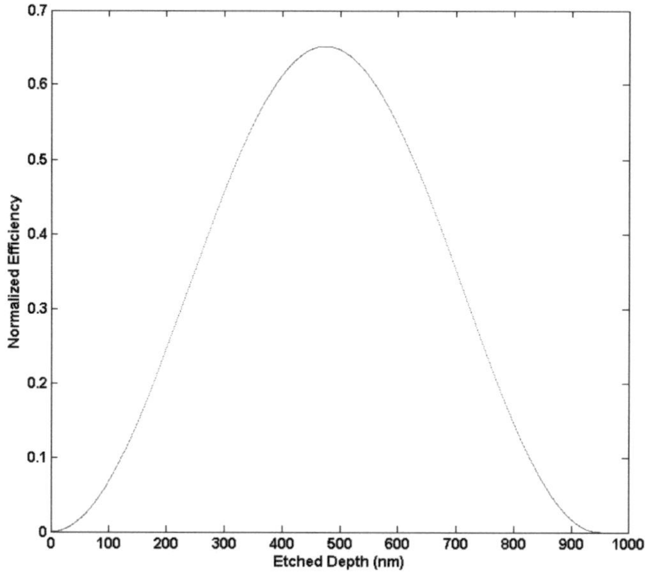

Fig. 24. Etched Depth against efficiency with photolithography.

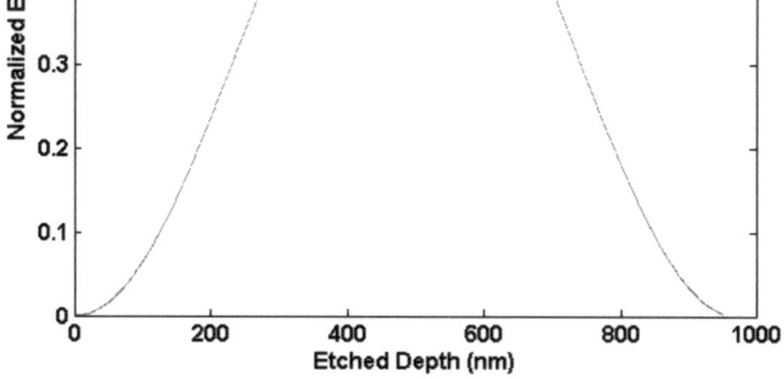

Fig. 25. Etched Depth against efficiency with EBL.

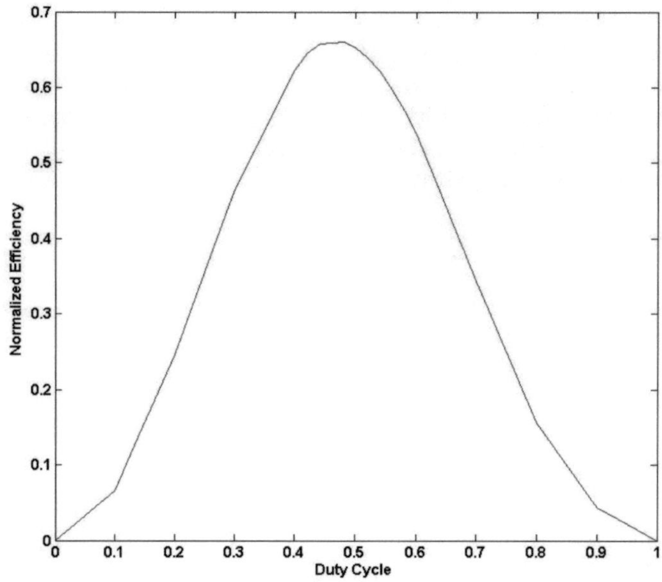

Fig. 26. Duty Cycle against efficiency.

Although fabrication errors on these structures will redistribute the energy among diffraction orders and increase the efficiency of central order, the total energy remains the same. It is also noted that the choice of fabrication method introduce different errors. For example, EBL tends to have variations in exposure dosage which has excess etch depth and shape errors. But it can provide very fine width and vertical shape walls and hence it can support high density structure. Conversely, chemical etching technique can provide very sharp side walls and accurate etch depth with carful timing. But, it never supports high density structure as the linewidth it can sustain is in order of um. The users must to balance between the cost and performance requirements.

5. Acknowledgement

This project is supported by GRF (No. 112710) of the Research Grant Council in Hong Kong.

6. References

[1] E. Hecht (2002). *Optics*, Addison Wesley, ISBN 0-80-538566-5, US

[2] C. B. Burckhardt (1966). Diffraction of a plane wave at a sinusoidally stratified dielectric grating. *J. Opt. Soc. Am.* Vol. 56 pp.1502–1509, ISSN 0030-3941

[3] M. G. Moharam & T. K. Gaylord (1982). Diffraction Analysis of Dielectric Surface Relief Gratings. *Journal of the Optical Society of America,* Vol. 72, pp.1385-1392, ISSN 0030-3941

[4] H. Dammann & K. Gortler (1971). High-efficiency in line multiple imaging by means of multiple phase holograms. *Optics Communications,* Vol.3, pp.312-315, ISSN 0030-4018

[5] H. Dammann & E. Klotz (1977). Coherent optical generation and inspection of two-dimensional periodic structures. *Optica Acta,* Vol.4, pp.505-515, ISSN 0950-0340
[6] J. Jahns *et al* (1989). Dammann gratings for laser beam shaping. *Optical Engineering,* Vol.28, pp.1267-1275, ISSN 0091-3286
[7] J. Turunen *et al* (1990). Optimization and fabrication of grating beam splitters. *Journal of Physics D: Applied Physics,* Vol.21, pp.102-105, ISSN 0022-3727
[8] J. F. Wen & P. S. Chung (2007). 2D optical splitters with Polymer Optical Fiber Arrays. *Journal of Optics A: Pure and Applied Optics,* Vol.9, pp.723-727 , ISSN 1464-4258
[9] F. J. Wen & P. S. Chung (2007). Two-dimensional optical splitters with polymer optical fibre arrays. *Journal of Optics A: Pure and Applied Optics,* Vol.9, pp.723-727, ISSN 1464-4258
[10] J. Jia, C. Zhou & L. Liu (2003). Superresolution technology for reduction of the far-field diffraction spot size in the laser free space communication system. *Optics Communications,* Vol.228, pp.271-278, ISSN 0030-4018
[11] S. Zhao, J. F. Wen & P. S. Chung (2007). Simple focal length measurement technique with a circular Dammann Grating. *Applied Optics,* Vol.46, pp.44-49, ISSN 0003-6935
[12] S. Zhao & P. S. Chung (2007). Collimation testing using a circular Dammann grating. *Optics Communications,*Vol.279, pp.1-6, ISSN 0030-4018
[13] C. Wu *et al* (1991). Optically pumped surface-emitting DFB GalnAsP/InP lasers with circular grating. *Electronics Letters,* Vol.27, pp.1819-1821, ISSN 0013-5194
[14] T. Erdogan *et al* (1992). Circularly-symmetric operation of a concentric-circle grating surface-emitting AlGaAs/GaAs quantum-well semiconductor laser. *Applied Physics Letters,* Vol.60, pp.1921-1923, ISSN 0003-6935
[15] C. Zhou, J. Jia & L. Liu (2003). Circular Dammann Grating. *Optics Letters,* Vol.28, pp.2174-2176, ISSN 0146-9592
[16] S. Zhao & P. S. Chung (2007). Design of Circular Dammann Grating, *Optics Letters,* Vol.31, pp.2387-2389, ISSN 0146-9592
[17] J. F. Wen, S. Y. Law & P. S. Chung (2007). Design of Circular Dammann Grating by employing the Circular Spot Rotation Method. *Applied Optics*, Vol.46, pp.5452-5455, ISSN 0003-6935
[18] J. F. Wen & P. S. Chung (2008). A new Circular Dammann Grating Using Hankel Transform. *Journal Optics A: Pure and Applied Optics*, Vol.10, 075306, ISSN 1464-4258
[19] J. F. Wen & P. S. Chung (2008). The use of Circular Dammann Grating for angle measurement. *Applied Optics*, Vol.47, pp.5197-5200, ISSN 0003-6935
[20] J. F. Wen, Z. Y. Chen, & P. S. Chung (2010). Area measurement at long-distance using a circular Dammann grating. *Applied Optics*, Vol.49, pp.648-652, ISSN 0003-6935
[21] J. F. Wen, Z. Y. Chen & P. S. Chung (2008). A Novel Distance Measurement Technique based on Optical Fractional Fourier Transform, *Proceedings of 13th Opto-Electronics and Communications Conference*, P-42, ISBN 978-0-85825-807-5, Syndey, Australia, July 7-10, 2008
[22] C. Zhou & L. Liu (1995). Numerical study of Dammann array illuminators. *Applied Optics*, Vol.34, pp.5961-5969, ISSN 0003-6935
[23] F. Bowman (1958). *Introduction to Bessel functions*. Dover, ISBN 0486604624, New York, USA
[24] J. Turunen et al (1990). Optimization and fabrication of grating beam splitters. *Journal of Physics D*, Vol. 21, pp.102-105, ISSN 0022-3727

[25] Masato Okano et al (2004). Optimization of Diffraction Grating Profiles in Fabrication by Electron-Beam Lithography. *Appl. Opt.*, Vol. 43, pp. 5137-5142, ISSN 0003-6935
[26] M. T. Gale (1997). Replication, In: *Binary optics fabrication in Micro-optics: Elements, Systems and Applications,* H. P. Herzig (Ed.), Taylor and Francis, ISBN 0748404813, London, England
[27] J. M. Miller et al (1993). Multilevel grating array generators: fabrication error analysis and experiments. *Applied Optics*, Vol.32, pp.2519-2525, ISSN 0003-6935
[28] C. Zheng (2005). *Micro-Nanofabricaton Technologies and Applications*, Springer, ISBN 7-04-017663-7, China

7

Fabrication of Binary Diffractive Lens on Optical Films by Electron Beam Lithography

Atsushi Motogaito and Kazumasa Hiramatsu
Graduate School of Engineering, Mie University
The Center of Ultimate Technology on nano-Electronics, Mie University
Japan

1. Introduction

Two types of lenses can focus light: an optical lens using refraction phenomenon and a diffractive lens using diffraction phenomena. Table 1 shows the characteristics of each lens. The focal length of the diffractive lens is controlled by the structures of the lens, as mentioned in detail in Section 2.2. This suggests that the focal length of the diffractive lens is independent of refractive index and curvature. Thus, application of diffractive lenses to UV optical elements or thin optical elements is possible.

	Refractive lens	Diffractive lens
Principle	Refraction	Diffraction
Structure	 Radius of curvature R	Close to wavelength 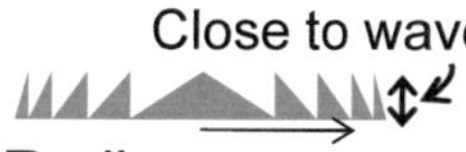 Radius r_m
Focal length f	$f = \dfrac{R}{2(n-1)}$	$f = \dfrac{r_m{}^2 - (m\lambda)^2}{2m\lambda}$
Control of f	n, R (depending on material)	r_m (m=1,2,···)
Short wavelength light such as UV and X-ray	Difficult	Possible (depending on material)
Thin structure	Difficult	Possible (wavelength order)
Controlling light distribution	Difficult	Possible

Table 1. Comparison of characters between refractive lens and diffractive lens

Recently, the emitting efficiency of light emitting diodes (LEDs) has improved; thus, they are used in lighting devices. To this end, miniaturizing the LEDs for smaller lighting devices and controlling the luminosity of LEDs are required. The conventional oval lamp-type LEDs cannot realize these requirements because the lens height of such LEDs is approximately 5 mm and its distribution of luminosity is determined by its shape. In this study, instead of the oval lamp-type lens, we used the diffractive lens on the optical films, as shown in Fig. 1. If the diffractive lens with short focal length (order of micrometer) can be fabricated, miniaturization of the lens system and consequently the LED lighting devices can be achieved. In order to realize the refractive lenses with the short focal length, large curvature radius is needed, thus making it difficult to realize it easily. Therefore, the diffractive lenses are suitable for realizing the short focal length lenses. Furthermore, by modifying the structure of the diffractive lens, it is easy to control the luminosity of the LEDs and the far-field pattern. Therefore, in this study, we focused on the diffractive lens because it enabled us to reduce the thickness of the lens, control the luminosity distribution of LEDs, and facilitate the realization of the binary structure.

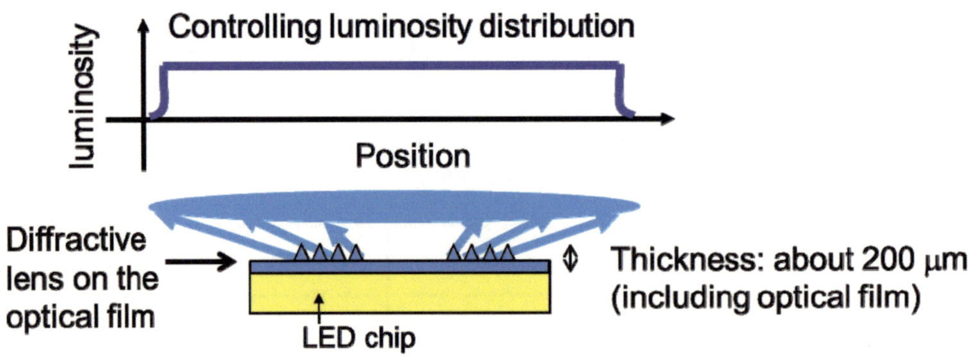

Fig. 1. Schematic representation of diffractive lens on the optical film

The zone plate was the first diffractive lens invented by I. L. Solet in 1875. To improve light efficiency, kinoform was invented by J. A. Jordan (Jordan et al., 1970). Recently, binary optics technology was developed using CAD design and VLSI technology (Swason and Veldkamp, 1989). The diffractive optical elements with multi-level grating having step-like cross-section have been developed. By controlling thestructure of the multi-level gratings, an optical effect almost same as that of the kinoform can be obtained (Orihara et al., 2001 & Yamada et al., 2004).

On the other hand, subwavelength structures (SWSs), which are equivalent to a blazed structure, were suggested by P. Lalanne (Lalanne et al., 1999 & Mait et al., 1999). These structures are fabricated binary SWSs converted from Fresnel lenses. These structures are fabricated easily than those of the multi-level gratings because they can be fabricated by electron beam lithography (EBL) or nanoimprint lithography (NIL). Furthermore, in the case of photolithography, combining some masks is not necessary. By using these structures, achromatized diffractive lenses were reported (Kleemann et al., 2008).

We aim to realize a highly effective short focal length diffractive lens using the binary diffractive lens fabricated by EBL, and expect the equivalent effect with the diffractive lens of the saw-like structure. NIL or photolithography can easily fabricate these structures at low cost and over a large area. However, EBL does not use any molds or masks. Therefore, it is convenient to examine EBL in detail to obtain optimum structures.

In this study, we carry out the design and fabrication of the binary diffractive lens with 2-mm focal lengths for controlling the luminosity distribution and the binary diffractive lens with the 100-µm-order focal length. Furthermore, to improve the diffraction efficiency, we characterize the detailed relationship between the lens structure and the light intensity.

2. Experimental procedure

In this section, the methods of design, fabrication, and characterization of the binary diffractive lens are described.

2.1 Basic optical characteristics of materials related to binary diffractive lens

The binary diffractive lenses, on which this study is focused, were fabricated on the poly(ethylene terephthalate) (PET) films. The PET films are often used as optical sheets for liquid crystal displays. There are many types of optical films such as polycarbonate (PC) and poly(methyl methacrylate) (PMMA). In this study, the EBL process was used for fabricating the binary diffractive lenses; this process required the optical films to endure high temperature and chemicals, making them more suitable than PC or PMMA.

In this study, the binary diffractive lens was fabricated by developing the resist for EBL (ZEP-520A, ZEON Co.) on the PET films (Teijin® Tetoron® Film, Teijin DuPont Films, Japan). If the refractive indexes of both materials are almost same, the binary diffractive lens can be fabricated by developing the resist instead of etching the PET films. Therefore, the refractive indexes of the PET films and the resist are evaluated by ellipsometry (M-2000DI, J.A. Woollam Co., Inc.). Fig. 2 shows the wavelength dispersion of the PET film and the resist on the PET film, including the data from the catalog of ZEP-520A. D2 and halogen lamps were used for this measurement. The refractive index of the PET film is relatively

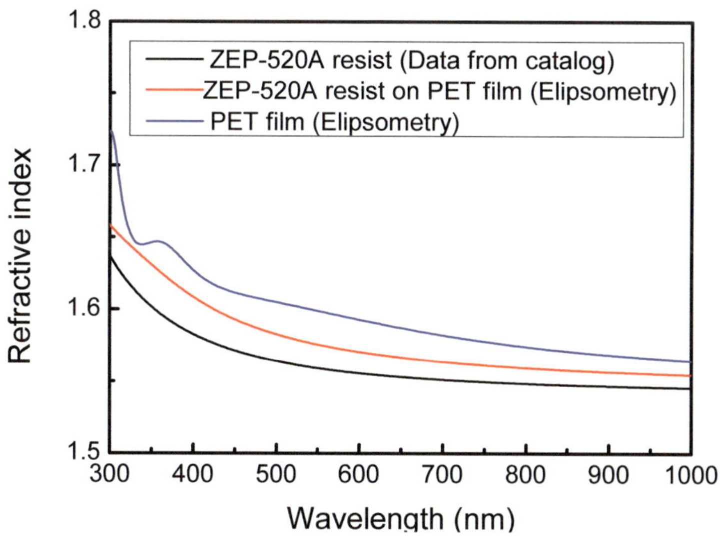

Fig. 2. Wavelength dispersion of the PET film and the resist on the PET film, including the data from the catalog of ZEP-520A

higher than that of the resist; however, in the visible region, their refractive indexes are between 1.58 and 1.60. Thus, in this study, these values are considered to be almost same. Furthermore, by using ellipsometry, the thickness of the resist is estimated using the multilayer model. Fig. 3 shows the relationship between the thickness of the resist and the number of rotations of the spin coater. The thickness of the resist varies from 760 to 460 nm and increases with the number of rotations. Thus, in this study, the binary diffractive lens structures of the electron beam (EB) resist were fabricated by developing an EB resist on the PET films. The development of the EB resist can be regarded as processing the surface of a PET film. The thickness of the resist is equivalent to the height of the binary diffractive lens.

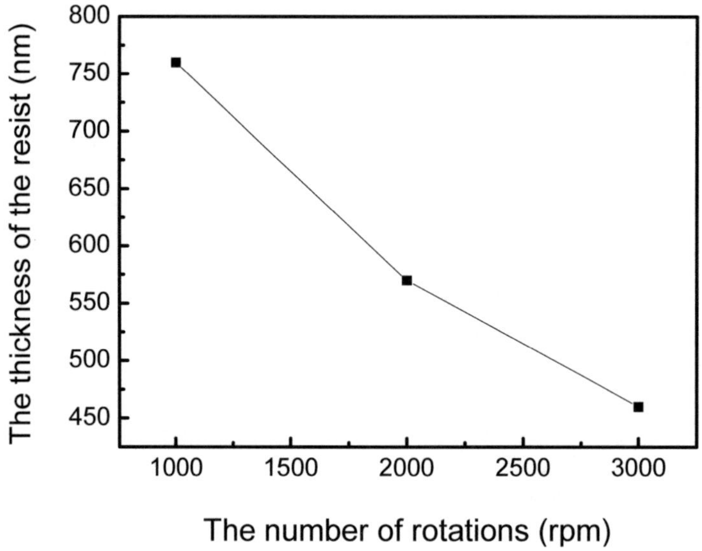

Fig. 3. Relationship between the thickness of the resist and the number of rotations of the spin coater

2.2 Design of diffractive lens

The fabricated binary diffractive lens was based on the micro-Fresnel lens. In this study, a part of two-level zone plates with a pattern of lines and spaces was fabricated. Radius of the mth zone r_m is

$$r_m = 2mf\lambda + (m\lambda)^2, \tag{1}$$

where f is the focal length of the designed lens and λ is the dominant wavelength. Equation 1 is based on the imaging theory of the diffractive lens (Buralli et al., 1989).

Then, mth period of this lens d_m is determined by $r_m - r_{m-1}$. In period d_m, the blazed structure is approximated to a step-like structure with n steps and then the step-like structures is converted to the relief structures by duty ratio of height $t_i = 1 - h(x_i)/h_{max}$ in each interval g, as shown in Fig. 4 (a). In the interval, the width of the air part is given by $g*t_i$. In the binary diffractive lens, N is the number of the relief structures in a period. Examples of the structures are shown in Fig. 4 (b) and the complete structure of this lens is shown in Fig. 4 (c).

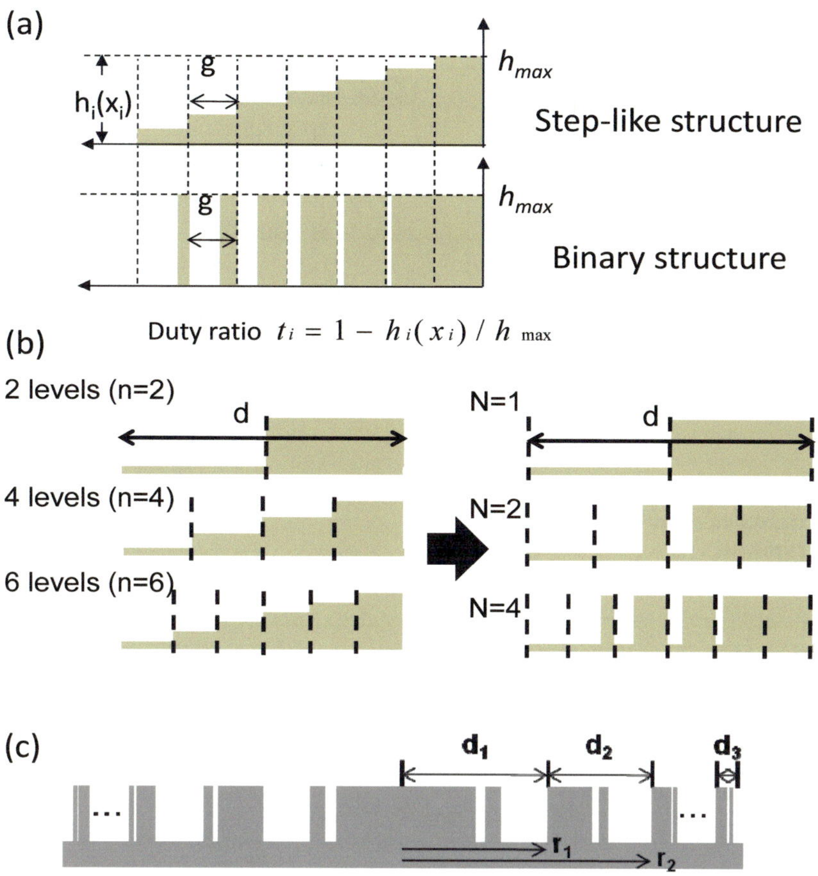

Fig. 4. Structure of the binary diffractive lens, (a) the conversion of the step-like structure to the binary structure, (b) the examples of the structures, (c) the complete structure

2.3 Fabrication of binary diffractive lens on optical films by EBL

The 125-μm-thick PET films were used as the substrate. Fig. 5 shows the procedure for the fabrication of the binary diffractive lens on the optical films by EBL. Before spin coating the EB resist, hexamethyldisilazane (HDMS) was spin coated on the surface of the PET film to improve the adherence between the PET film and the EB resist (Fig. 5 (a)). The surface was spin coated with an EB positive resist followed by pre-baking (Fig. 5 (b)). Then, the charge-up prevention was spin coated on the EB resist (Fig. 5 (c)).

The EBL system (Crestec CABL-8000) was equipped with a ZrO/W thermal field emission cathode. The acceleration voltage was 30 kV; the electrons accelerated by this voltage were able to penetrate the resist (Fig. 5 (d)). After exposure, the resist was developed and the binary diffractive lens could be obtained from these procedures (Fig. 5 (e)). The size of the patterns for the binary diffractive lens ranged from 100 × 100 μm² to 2 × 2 mm².

The optimum results obtained using the diffractive lenses fabricated by EBL, such as period, width, and height of the fabricated binary diffractive lenses, are useful for fabricating the molds of the thermal-type nanoimprint.

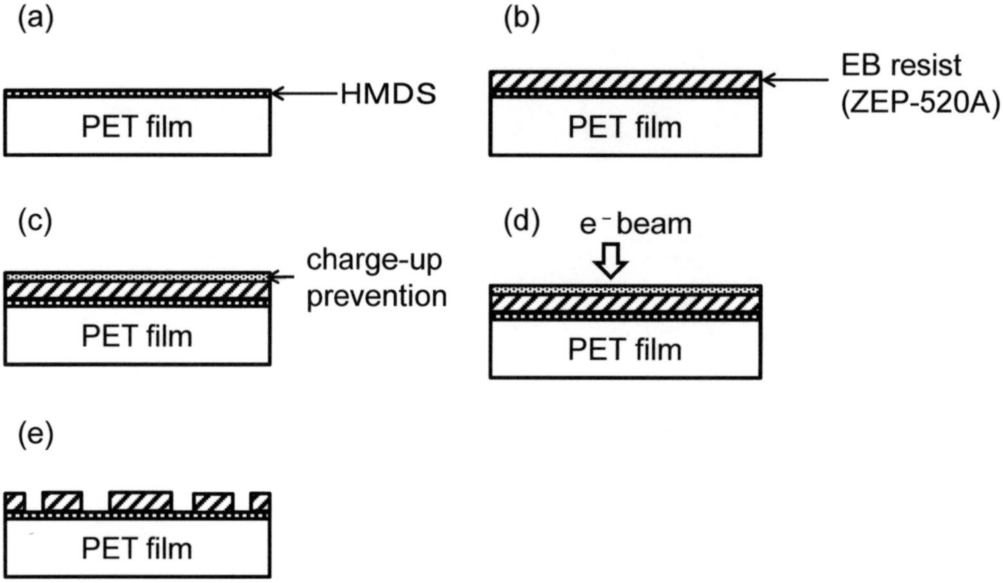

Fig. 5. Procedure for the fabrication of the binary diffractive lens on optical films by EBL, (a) spin coating HMDS, (b) spin coating EB resist and pre-baking, (c) spin coating charge-up prevention, (d) exposing e^- beam, (e) developing the resist and obtaining the binary diffractive lens

3. Results and discussion

In this section, we describe and discuss the experimental results. There are two types of the binary diffractive lenses: (1) the binary diffractive convex lens with a 2-mm focal length for controlling the luminosity of LED light and (2) the binary diffractive convex lens with a 150-μm focal length.

3.1 Binary diffractive convex lens with 2-mm focal length for controlling luminosity of LED light

The binary diffractive convex lens with 2-mm focal length was fabricated on the PET film. Fig. 6 shows the scanning electron microscopy (SEM) image of the fabricated binary diffractive lens on the PET film. The diffractive lens having width almost same as that of the designed lens was obtained.

Optical characterization of the fabricated binary diffractive lens was carried out. The luminous intensity distribution of the LED (λ = 566 nm) for the binary diffractive lens was characterized using a luminous intensity distribution system (Asahi Spectra IMS5000-LED).

The fabricated lens was then mounted on the LED chip and spectral irradiance in the vertical direction was measured; Fig. 7 shows the distribution of the irradiance. Most of the LED light was focused, as shown in Fig. 7 (a); the light distribution angle became narrow (30°) using the binary diffractive lens. As shown in Fig. 7 (b), spectral irradiance around 0° with this lens was 1.5 times higher than that without the lens. On the other hand, two side peaks in these data were observed and believed to be due to light escaping from the fabricated binary diffractive lens. From these results, it is clear that the luminous intensity distribution can be controlled using this type of lens.

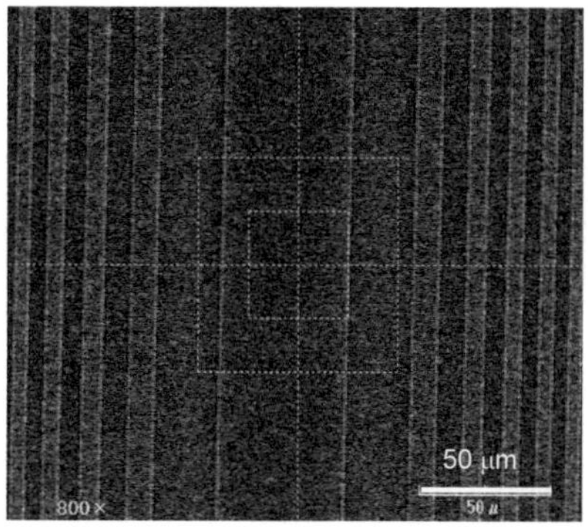

Fig. 6. SEM image of the fabricated binary diffractive convex lens with 2-mm focal length on the PET film.

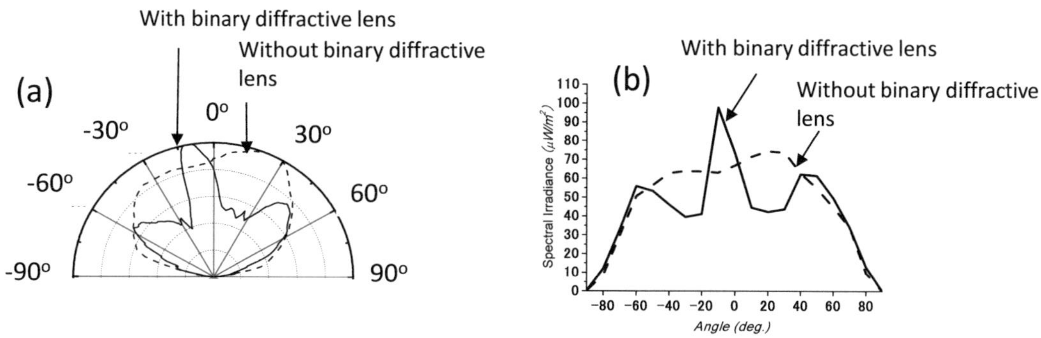

Fig.7 Ddistribution of the irradiance. (a) Angle dependence of normalized spectral irradiance. (b) Angle dependence of the absolute value of spectral irradiance

3.2 Binary diffractive convex lens with 150-μm focal length

Although the binary diffractive lens was effective in controlling the luminous intensity, diffraction efficiency was reduced when the diffraction angle was decreased (Lalanne et al., 1999; Kleemann et al., 2008). Furthermore, the focal length of the fabricated binary diffractive lens is 2 mm. In order to realize a thin LED light source, the focal length has to be shorter. In this section, to improve the diffraction efficiency and shorten the focal length, we designed the binary diffractive convex lens with 150-μm focal length.

In this study, a binary diffractive lens with a focal length of approximately 150 μm was designed and light propagation of the plane wave was simulated by the finite domain time difference (FDTD) method. Fig. 8 shows the field intensity distributions for TE polarization of the binary diffractive lens. The simulation parameters were λ = 632 nm, n = 1.575 (refractive index of the PET film), and n_0 = 1.0 (refractive index of air). The value of the period in part of the fringe was smaller than that in the center. The designed lens was placed along the x-axis (z = 0). The light was incident from z = 0 to the +z direction, resulting in the light being focused at x = 0 μm and z = 140 μm. After focusing, the light was spread with time because of diffraction. Therefore, a binary diffractive lens with a micrometer-order focal wavelength is expected to provide a small and thin light source for controlling the luminous intensity distribution. On the basis of the results of section 3.1, we speculated that the LED light can be focused at 140 μm.

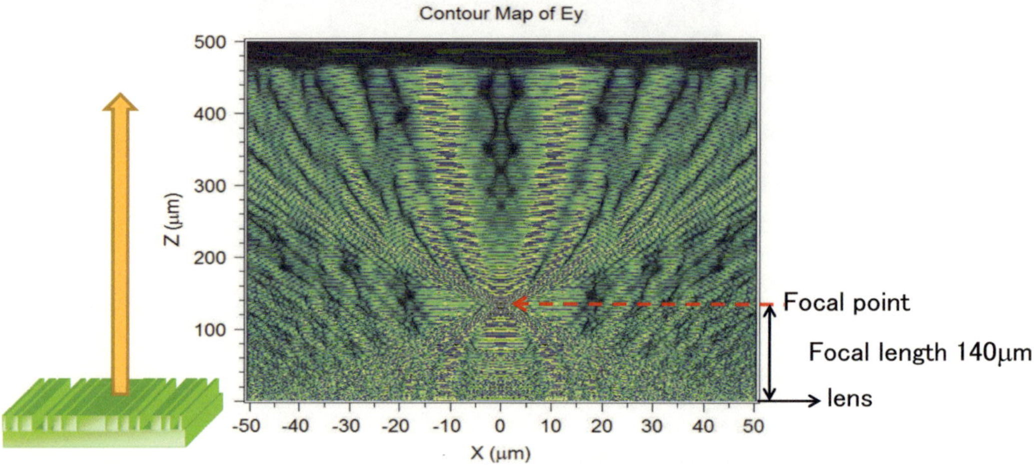

Fig. 8. Field intensity distributions for TE polarization of the binary diffractive lens

The binary diffractive lens with a 150-μm focal length was fabricated; its size was 100 × 100 μm^2 and thickness was 570 nm, as measured by ellipsometry. Fig. 9 shows the SEM image of the fabricated binary diffractive lens (N = 4) on the PET film. The diffractive lens, whose width was almost the same as the designed lens, was obtained.

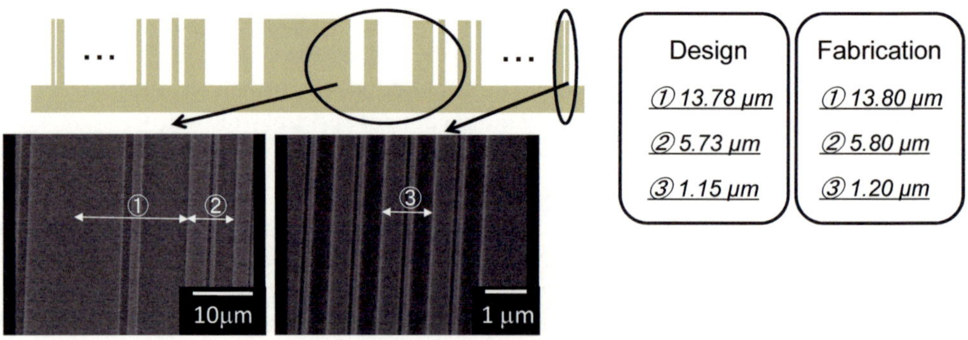

Fig. 9. SEM image of the fabricated binary convex diffractive lens with a 150-μm focal length (N = 4) on the PET film

The far-field transmitted intensity distribution of the fabricated lens is characterized by red laser light (λ = 635 nm). The aperture with a diameter of 100 μm was used for eliminating the light escaping from the edge of the lens. Fig. 10 shows the far-field transmitted intensity distribution of the fabricated lens with different N values (1, 2, 4). The focal length of this lens, which is estimated from this distribution, is approximately 160 μm, which is almost same as that in the FDTD simulation. For higher N values, the intensity of first-order diffraction decreases.

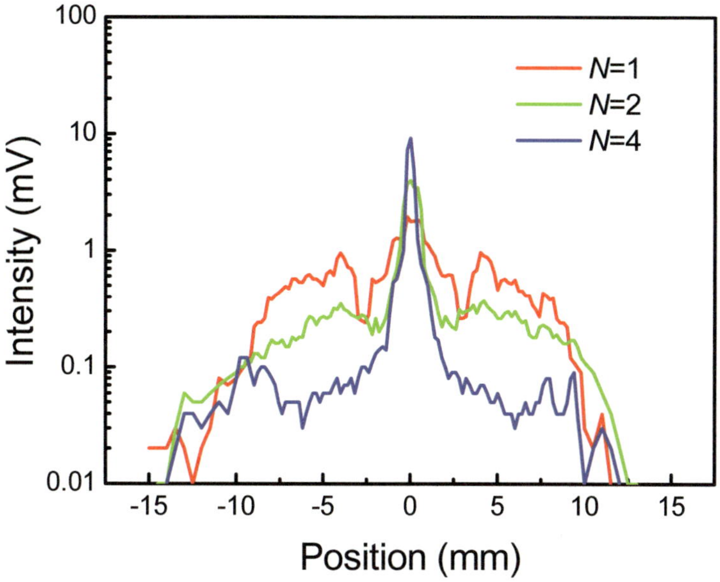

Fig. 10. Far-field transmitted intensity distribution of the fabricated lens

To determine the reason for these results, the binary diffractive lenses with only first period (d_1 = 13.78 μm) and 12th period (d_{12} = 2.12 μm) were fabricated. Fig. 11 shows the far-field light distribution of both lenses. In the case of d_1 = 13.78 μm, the first-order diffraction is observed when N = 4. Because d_1 is considerably larger than the wavelength of light, the first-order diffraction cannot be observed when N is small. On the other hand, in the case of d_{12} = 2.12 μm, the first-order diffraction is observed when N = 1, and it disappears by increasing the number of N. Therefore, in order to improve the diffraction efficiency of the diffractive lens, it is necessary to control the intensities of the zero- and first-order diffractions by choosing the binary structures.

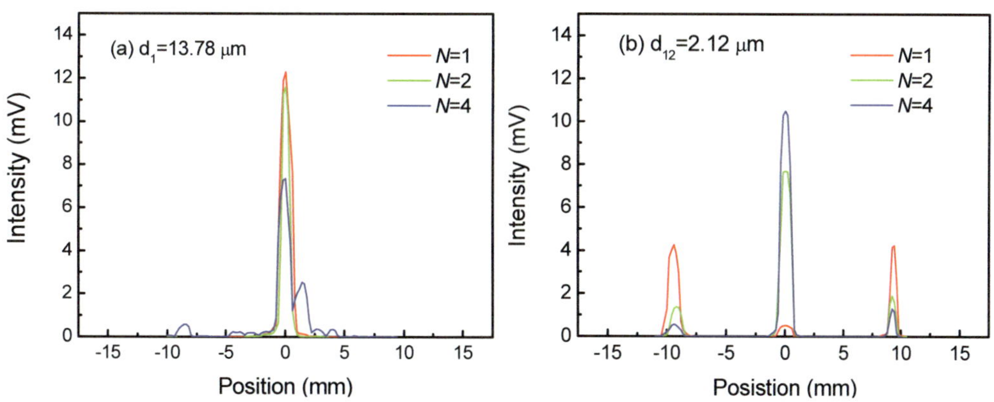

Fig. 11. Far-field light distribution of the binary diffractive lenses in the case of (a) d_1 = 13.78 μm and (b) d_{12} = 2.12 μm

4. Conclusion

In summary, we designed and fabricated two types of binary diffractive convex lenses using EBL on a PET film. In the case of the binary diffractive convex lens with 2-mm focal length, it is possible to control the luminous intensity distribution. To improve the diffraction efficiency and realize a thin LED light source, we designed a binary diffractive lens with 140-μm focal length. This type of lens with focal wavelengths in the micrometer range can produce a thin LED light source to control the luminous intensity distribution.

To realize the binary diffractive lens with the 100-μm-order focal lengths, we characterize the relationship between the diffractive lens structure and its light intensity. It is clear that the intensities of the zero- and first-order diffractions are controlled by the structure of the binary diffractive lens. By using this lens, wide luminous intensity distribution can be obtained.

5. Acknowledgment

This work was partly supported by a Grant-in-Aid for Scientific Research from the Japan Society for the Promotion of Science (No. 18360008 and 21360007), Mie University COE Projects, a Start-up Grant Program for Mie Venture Business from Mie Industry Enterprise Support Center, the Kinki Invention Center, and the Knowledge Cluster Initiative from the Ministry of Education, Culture, Sports, Science and Technology.

The authors thank Mr. T. Sato (JA Woollam Japan Co., Inc.) for ellipsometry measurement.

The authors also thank Dr. H. Miyake, Mr. K. Manabe, Mr. N. Machida, Mr. Y. Nakayama, Mr. K. Arakawa, and Mr. Y. Seriguchi for their help in the experiments and valuable discussions.

6. References

Jordan, J. A. ; Hirsch, Jr. P. M. ; Lesem, L. B. & Van Rooy D. L. (1970). Kinoform Lenses. *Applied Optics*, 9, 8, (1883-1887), ISSN0003-6935

Swason, G. J. & Veldkamp, W. B. (1989). Diffractive optical elements for use in infrared systems. *Optical Engineering*, 28, 6, (605-608) ISSN0091-3286

Orihara, Y. ; Klaus, W. ; Fujino, M. & Kodate, K. (2001). Optimization and Application of Hybrid-Level Binary Zone Plates. *Applied Optics*, 40, 32, (5877-5885), ISSN0003-6935

Yamada, K. ; Watanabe, W. ; Li, Y. ; Itoh, K. & Nishii, J. (2004). Multilevel Phase-type Diffractive Lenses in Silica Glass Induced by Filamentation of Femtosecond Laser Pulses. *Optics Letters*, 29, 16, (1846-1848), ISSN0146-9592

Lalanne, P. ; Astilean, S. ; Chavel, P. ; Cambril E. ; Cambril, E. & Launois, H. (1999). Design and Fabrication of Blazed Binary Diffractive Elements with Sampling Periods Smaller than the Structural Cutoff. *Journal of Optical Society of America. A*, 16, 5, (1143-1156) ISSN0740-3232

Lalanne, P. (1999). Waveguiding in Blazed-Binary Diffractive Elements. *Journal of Optical Society of America. A*, 16, 10, (2517-2520) ISSN0740-3232

Mait, J. M. ; Prather, D. W. & Mirotznik, M. S. (1999). Design of Binary Subwavelength Diffractive Lenses by Use of Zeroth-order effective-medium Theory. *Journal of Optical Society of America. A*, 16, 5, (1157-1167) ISSN0740-3232

Kleemann, B. H. ; Seesselberg, M. & Ruoff, J. (2008). Design Concepts for Broadband High-efficiency DOEs. *Journal of the European Optical Society*, 3, (08015-1–08015-16) ISSN 1990-2573

8

Photocontrolled Reversible Dimensional Changes of Microstructured Photochromic Polymers

Despina Fragouli, Roberto Cingolani and Athanassia Athanassiou
Center for Biomolecular Nanotechnologies @UNILE, Istituto Italiano di Tecnologia Italy

1. Introduction

Stimuli-responsive polymeric materials are able to change their chemistry and their conformation upon an external signal. The external signal may be derived from a change in temperature, chemical composition or applied mechanical force of the specific material, or can be triggered externally with exposure to an electric or magnetic field or to light irradiation. In this respect, a photochromic substance is a stimuli responsive material which is characterized by its ability to alternate between two different chemical forms having different absorption spectra, in response to light irradiation of appropriate wavelengths (Brown 1971). Due to this important property, a significant amount of effort has been devoted to the formation of polymeric materials functionalized with photochromic molecules for the creation of photosensitive "smart material" systems, that change reversibly their physical and chemical properties by the use of light. The corresponding reversible effects of the molecules such as dipole moment, surface energy, refractive index, and volume are preserved in the polymer matrix, and have numerous promising applications in devices for three-dimensional (3D) optical memories, (S. Kawata & Y. Kawata 2000), in actuators (Yu et al 2003, Athanassiou et al 2005), in holographic or diffractive optics, (Fu et al 2005, Tong et al 2005) or in microfluidics, (Caprioli et al 2007, Walsh et al 2010) etc. Concerning microfluidic devices using photochromic plastic films, the transportation of fluids happens without the need for their molecules to be charged, as done in other studies (Mitchel 2001). This is achieved by gradually modifying the surface tension, and thus the wettability, by irradiating with increasing time along the direction of the fluid movement (Ichimura et al 2000). The gradual wettability changes are exclusively based on the photochemical modification of the embedded photochromic molecules caused by the photoisomerization process. In addition, in the case of the diffraction gratings the development was generally done by interference of different polarized laser beams, or by electric-field application, and the modification of their diffraction efficiency is connected with the changes of the refractive index of the photochromic molecules during this procedure (Yamamoto et al 2001, Fu et al 2005).

Here we present how the volume changes induced to the photochromic polymers by the photoisomerization of their embedded photochromic molecules, can improve significantly the performance of these two different type of applications, namely the microfluidic devices and the diffraction gratings. In both cases the lithographic technique used for the microstructuring of the photochromic doped plastic films is the soft molding. Concerning the microfluidics applications, the presented microstructured photochromic plastic films exhibit a significant improvement on the reversible wetting characteristics compared to those on the flat surfaces. This improvement is due to the combination of the changes in the surface polarity and thus in the wetting properties with the modified surface conformation, both provoked by the light induced changes of the photochromic molecules. Moreover, regarding the optical gratings, we present a different approach in where the control of their diffraction efficiency relies on the dimensional variations of the gratings upon laser irradiation. Following this approach, the efficiency of the gratings is significantly improved with respect to previous works.

Such findings open the way for the production of optically switchable gratings based on reversible dimensional changes, and can be of great importance in all-optical signal processing systems. Moreover, the ability to control the wettability of surfaces by microstructuring and to tune it by using photochromic molecules, permits the application of these lithographically formed structures to all-optically controlled switches capable of operating with tunable speed, and to microfluidic actuation.

2. Materials and methods

2.1 Photochromic polymers and properties

Photochromic doped polymer films were prepared by incorporating photochromic molecules into polymer matrices. In particular, solutions of the polyethylmethacrylate-co-methylacrylate copolymer (PEMMA) (average molecular weight, M_w=100,000) (Aldrich) mixed with the photochromic molecule 1′,3′-dihydro-1′,3′,3′- trimethyl- 6-nitrospiro[2H-1-benzopyran-2,2′-(2H)-indole] or (6-NO_2 BIPS) (Aldrich) (weight ratio 90/10 respectively or 95/5) were prepared in toluene. Consequently a certain volume of this solution was spin-coated onto a glass substrate.

The photochromic dopant 6-NO_2 BIPS, belongs to the family of spiropyrans (SP) which have been extensively studied in the past decades (Görner 1998). Initially it has a 3D structure, and exists predominately in its non polar form (Figure 1). It is colorless, denoting that it is transparent in the visible range of the spectrum, but absorbs in the ultraviolet (UV). Upon irradiation with UV light, it is converted to its isomeric form, merocyanine (MC), through the photochemical cleavage of its carbon–oxygen (C_{spiro}-O) bond. The MC has a planar structure, it is colored and polar, and has a new absorption band in the visible range of the spectrum. MC can revert back to the SP form photochemically, using visible-light irradiation.

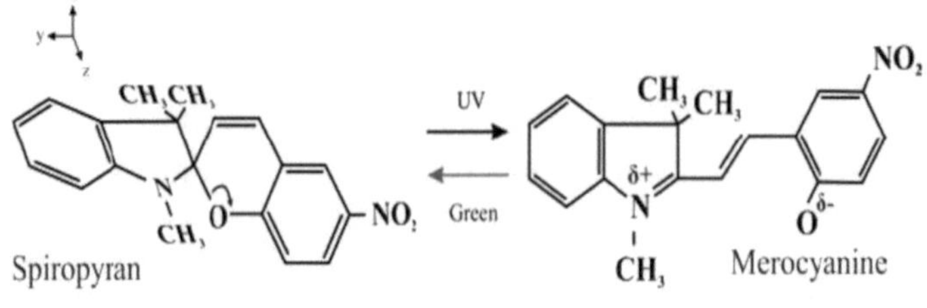

Fig. 1. The photochromic dopant spiropyran (SP) and the stable form of its isomer merocyanine (MC).

Briefly, UV and visible irradiation causes the reversible transformation of these chemical species, between two states (isomers) that have light absorption bands in characteristic spectral regions. This property is retained when the photochromic molecules are incorporated in polymer matrices, where they are homogeneously dispersed forming miscible systems. Specifically, the absorption properties of the photochromic polymer films prepared as described above change reversely upon UV-visible irradiation as shown at Figure 2. Initially the system is transparent at the visible range of the spectrum. Upon pulsed UV laser irradiation the SP is slowly converting to the MC isomer, fact indicated by the new absorption band in the visible region of the spectrum (ca. 565 nm). The intensity of the peak increases with the number of UV pulses until a plateau is reached, which suggests that the photoisomerization is completed and that the system has reached the equilibrium. The subsequent irradiation with green laser light, causes the decrease of the intensity of the previously formed MC peak, while after a certain number of pulses the spectrum reaches its initial form, indicating that MC reverts fully to the SP isomer. These data confirm that under the irradiation conditions mentioned in the figure caption of Figure 2, the reversible properties of SP are retained in the host polymer matrix. Depending on the irradiation conditions and the weight percentage of the photochromic molecules in the polymer matrix (usually ≤10%), it has been shown that typically about 4-10 irradiation cycles can be performed, while further irradiation causes the degradative photooxidation of the photochromic molecules, restricting thus the lifetime of the system (Athanassiou et al 2006c). Additionally, the degradative phenomena start to be evident usually after the third cycle. In order to exclude this parameter from the following study, results derived by the first three irradiation cycles are presented.

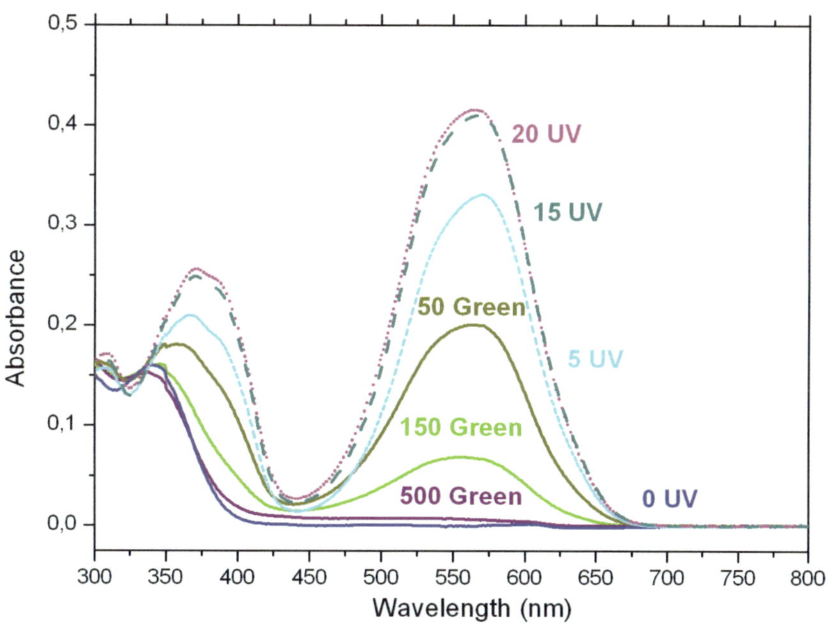

Fig. 2. Absorption spectra of the PEMMA/SP 10% wt upon UV and visible irradiation. For the specific study, the irradiation conditions used are: λ_{UV}=355 nm, fluence F_{UV}=20 mJ cm^{-2}, λ_{vis}=532 nm, F_{vis}=35 mJ cm^{-2}.

Except of the alteration in the optical properties, there are also other physical and chemical properties that change reversibly upon UV-visible irradiation even if the SP molecules are incorporated in the polymer matrix, such as dipole moment, surface energy, refractive index, and volume. Concerning the volume changes upon UV irradiation, occurs the formation of aggregates between different MC stereoisomers with zwitterionic character causing density fluctuations in the polymer matrix, reducing thus the MC partial molar volume. Consequently, takes place the short scale motion of the polymer chains in order to diminish the density fluctuations in the samples, and this leads to the macroscopic reduction of the dimensions of the matrix. This effect is reversible, since upon green irradiation, MC molecules return to the SP form, which does not form aggregates (Athanassiou et al. 2005)

2.2 Substrates microstructuring: soft molding

For the microstructuring of the photochromic plastic films it is used the soft molding lithography (SM). It is actually based on the conformal contact between the material to be patterned and an elastomeric replica of a master structure, and it combines soft and nanoimprint lithography, using elastomeric elements and exploiting the glass transition of organic compounds. Particularly, an elastomeric mold is placed onto a polymeric film applying the pressure of its own weight, and consequently is heated up above the films' glass transition temperature, T_g. The subsequent cooling down, below T_g, freezes the pattern into the polymer, and the replica is peeled off. The micropatterns formation is based on the capillarity effect that drives the polymer to penetrate into the recessed features of the elastomeric replica. The SM presents various advantages compared to the nanoimprint lithography. Specifically, since penetration of the polymer into recessed features of the replica is driven by capillarity effects, SM is only slightly affected by problems caused by difficult polymer transport. Moreover, it does not need any pressing setup to ensure contact between the mold and the polymer. Finally, SM does not present pattern shrinkage and distortion due to the solvents employed by other soft lithography techniques. (Pisignano et al 2004)

In Figure 3 it is represented the process followed for the SM. Initially, the original master structures are fabricated onto glass or Si by both photo- and electron-beam lithography. The realized masters are used as templates on which elastomeric replicas were realized using polydimethylsiloxane (PDMS, Sylgard 184, Dow Corning, Midland, MI) according to a standard replica molding procedure, and placed onto polymer films under their own weight. (Pisignano et al 2004)

In order to form the microstructured photochromic polymer substrates for the wettability study, elastomeric molds of PDMS having periods α= 1.3, 28.0, and 180.0 μm were placed onto flat spin-cast films. Then the system was heated at 50 °C, a temperature higher than the T_g of the PEMMA films (T_g = 48 °C). After the thermal cycle, the replica was easily peeled off from the photochromic polymer substrates, on which the patterns are transferred. For the preparation of the gratings on the films for the diffraction efficiency study, the substrates were placed on a hot plate and heated until they reach a temperature $T_{grating}$ of 65 °C. Then an elastomeric mold with α=4 μm was placed on the substrate for 10 min, resulting in the formation of the gratings. In both cases the SM procedure was carried out in nitrogen atmosphere to avoid the deterioration of the photochromic molecules upon heating.

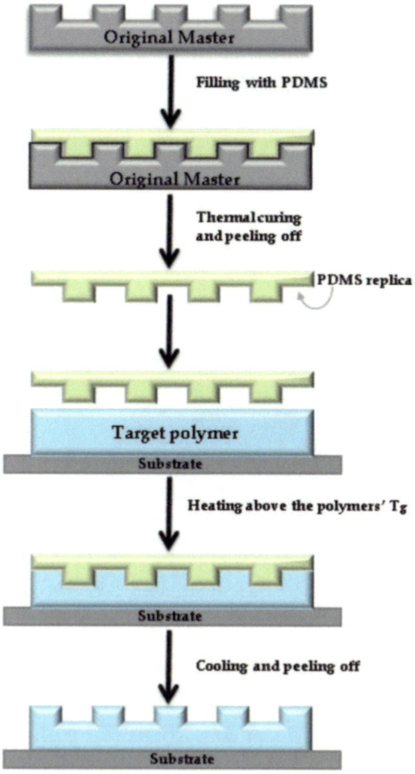

Fig. 3. Schematic diagram of the process of master replication and soft molding.

3. Light induced wettability changes of patterned substrates

3.1 Introduction

As abovementioned, the modification of the wetting characteristics of photochromic surfaces depends mainly on the photochemical processes which modify the surface tension and are caused by the UV-visible irradiation cycle. However, studies on patterned surfaces have demonstrated that the surface roughness affects significantly the wettability properties (Patankar 2003). Here it is studied the combined effect of the two aforesaid factors on the wettability properties of patterned photochromic polymeric surfaces. The wettability changes induced by photomechanical and photochemical changes are reversible upon UV-visible light irradiation, resulting in reversible changes of the wetting properties of the surfaces.

In particular, it is shown that the hydrophilicity of the photochromic polymeric surfaces is increased upon UV laser irradiation due to the polarity change caused by the photoisomerization, while the process is reversed upon green laser irradiation. The microstructuring of the surfaces enhances significantly the hydrophobicity of the system due to the increased surface roughness, and the light-induced wettability variations of the structured surfaces are enhanced by a factor of 3 compared to those on the flat surfaces. (Athanassiou et al 2006a, 2006b) In addition, by changing the topological parameters of the introduced pattern (e.g. by decreasing the period), are achieved higher differences in the surface wetting properties (Lygeraki et al 2008).

3.2 Results and discussion

Figure 4, shows the apparent water contact angle (WCA) of a drop on a flat photochromic polymer surface before and after pulsed UV and green laser irradiation. The WCA on the flat substrate, θ_Y, is formed when the liquid is in contact with a solid surface in static equilibrium with its vapor, and is determined by the Young's equation:

$$\gamma_{LV} \cos\theta_Y = \gamma_{SV} - \gamma_{SL} \tag{1}$$

where γ_{LV}, γ_{SV} and γ_{SL} represent the interfacial tensions at the boundaries between the liquid (L), vapor (V), and solid (S).

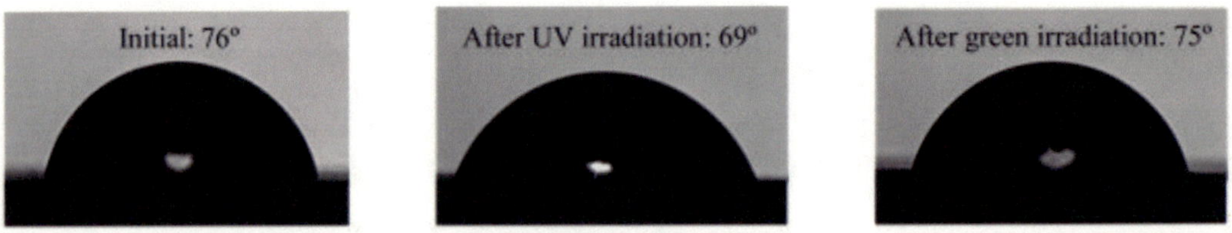

Fig. 4. WCA of a water drop of volume 3 μl on flat photochromic polymer surface before and after UV-green laser irradiation. (5% wt of SP in PEMMA, λ_{UV}=308 nm, F_{UV}=40 mJ cm^{-2}, λ_{green}=532 nm, F_{green}=45 mJ cm^{-2}) (Athanassiou et al 2006a)

As shown, before any irradiation the surface is hydrophilic with a WCA of 76°. After irradiation with enough UV laser pulses so as to reach the complete photoisomerization of all the non polar SP molecules to the polar MC form (50 pulses), the surface becomes more hydrophilic, with a WCA of 69°. The subsequent green irradiation with 500 laser pulses causes the reversible phenomenon, which is the conversion of the MC isomer to the SP form, and the increase of the WCA until it reaches its initial value. The maximum WCA difference measured on numerous flat surfaces upon UV irradiation was 7°±1°.

Fig. 5. Atomic force microscopy images of typical patterned surfaces of photochromic polymers. (Athanassiou et al 2006a)

In order to examine the effect of the photomechanical changes upon the same irradiation conditions, the aforementioned surfaces were microstructured using the SM technique. The replica used had a period α=1.3 μm, and by following the already described steps there were formed on the photochromic polymeric surface patterns of the same period (Figure 5). The WCA on the specific surface was found to be greatly affected by the patterning, showing an increased hydrophobicity with a value almost 30° greater than that of the flat surfaces (Figure 6). Moreover, always in comparison with the flat surfaces, the light induced WCA changes due to the photoisomerization effect are enhanced by a factor of 3, since the WCA change before and after UV irradiation is ca. 20° (WCA change on the flat surface, ca. 7°). This guides to the conclusion that the microstructuring affects significantly the reversible photoinduced wettability changes of the surfaces.

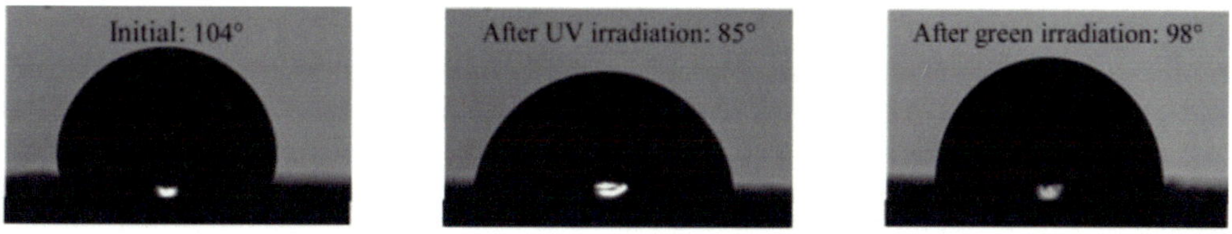

Fig. 6. WCA images obtained on patterned surfaces before and after laser irradiation. (5% wt of SP in PEMMA, λ_{UV}=308 nm, F_{UV}=40 mJ cm^{-2}, λ_{green}=532 nm, F_{green}=45 mJ cm^{-2}) (Athanassiou et al 2006a)

In order to explain the effect of roughness on the wetting characteristics of a surface, there are proposed two theories. The first is referred to as the Cassie-Baxter model (Cassie and Baxter 1944) (Figure 7a), and describes the wettability of rough surfaces, where only partial wetting may occur due to the trapping of air underneath the drop at the recessed regions of the surfaces. Since the drop is situated partially on air, the surface exhibits an enhanced hydrophobic behavior. The second one is the Wenzel model (Wenzel 1936), and it proposes that roughness increases the liquid-solid interfacial area, and thus hydrophilic surfaces (θ<90°) become more hydrophilic, and hydrophobic (θ>90°) more hydrophobic (Figure 7b).

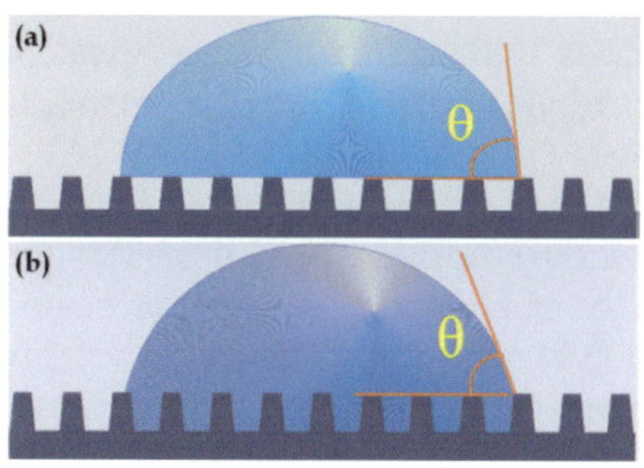

Fig. 7. Representation of a drop on a patterned surface, according to Cassie-Baxter model (a) and to Wenzel model (b).

In the presented cases the WCA of the drop on the flat surface is 76° thus hydrophilic (<90°) and according to the Wenzel model the patterning should increase its hydrophilicity. However, the experimental results presented above show that the WCA after the patterning is significantly increased, reaching a maximum value of 104°, indicating that the surface became hydrophobic (>90°). Thus, the presented surfaces follow the Cassie-Baxter model, where the relation with the WCA of the flat surface (θ_Y) is given by the following equation:

$$\cos\theta_r = -1 + f_s(1 + \cos\theta_Y) \qquad (2)$$

with f_s the solid fraction of the surface in contact with the liquid. According to this equation, the WCA decreases when the f_s is increasing. Using the WCA (θ_Y) measured on the flat surface and the one of the patterned in all three cases, before, after UV, and after green irradiation, the f_s is calculated. The results show that the fraction of the patterned surface in contact with the liquid is increased after UV irradiation (f_s=0.8) compared to the no irradiated sample (f_s=0.6). AFM microscopy studies of the topological changes of the gratings after UV irradiation, showed that the average volume decrease of each nanoimprinted stripe is ca. 30 nm. However, since the f_s is higher, and the stripes of the pattern are narrower, it is concluded that the water drop penetrates deeper into the channels of the UV irradiated pattern (partial wetting-Figure 8b), decreasing thus the WCA value. After irradiation with green laser pulses the MC molecules return to their SP isomers, the stripes recover their previous volume, and thus the drop returns to its previous condition, demonstrating once more that the wetting behavior is greatly influenced by both, the changes in the surface polarity and the volume of the stripes (Athanassiou et al 2006a).

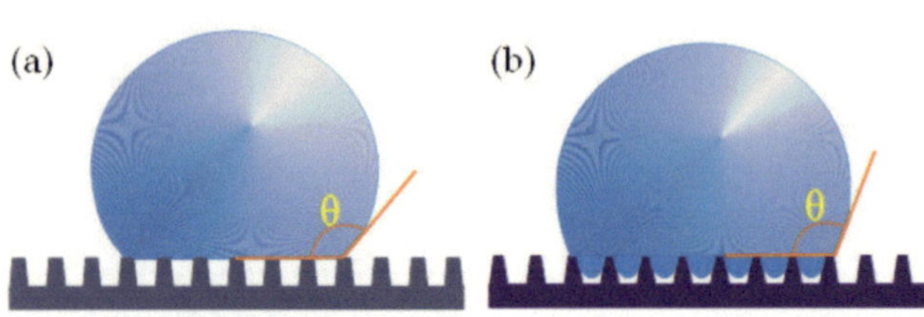

Fig. 8. Representation of a drop on a patterned surface, according to Cassie-Baxter model, before (a) and after (b) UV irradiation.

A further demonstration of the abovementioned statement is the study of the wetting properties of patterned surfaces with different topological parameters. In particular, the periods of the elastomeric replicas used for the formation of the patterns on the photochromic polymer surfaces (10% wt of SP in PEMMA) were 1.3, 28.0, and 180.0 μm. Figure 9a demonstrates the WCA changes upon three UV-green irradiation cycles for each of the patterned substrates and the comparison with the flat surface. For comparison reasons it is presented also the WCA of the pure polymeric surface, without any addition of photochromic molecules, which is not affected by the irradiation as expected. As shown, the patterning of the surfaces increases the initial WCA in all three cases. However, the smaller the period of the pattern the more hydrophobic is its behavior compared to the flat surface. After UV irradiation, all substrates become more hydrophilic, but again the WCA difference from the initial one is greater at the pattern with the smaller period, and becomes smaller as the period is increasing (Figure 9b).

The observed difference between the various patterns cannot be attributed to the surface chemistry changes upon UV-green irradiation, since this is the same in all samples. Additionally, if this was happening, it should be observed the inverse phenomenon, thus the greater change for the pattern with the greater period, since in these cases the drops are in contact with a greater percentage of the solid surface, as seen from the calculated value of f_s at Table 1. Thus, the reported differences can be attributed to the volume shrinkage of the patterns upon UV irradiation. This reduction of the patterns volume affects much more the contact angle of the samples with patterns of smaller period, since the drop lies on a greater number of patterned features. Indeed it is calculated that the radius of the drop is 20 times larger than the size of each feature projected on the plane of the interface in the case of the patterns with a 180 µm period, about 140 times greater in the case of the patterns with a 28 µm period, and finally more than 2950 times greater in the case of the patterns with a 1.3 µm period. Accordingly, as shown at the table, for the pattern with period of 180 µm there is a slight change of the *f* upon UV-green irradiation. This is not the case for the smaller pattern where the change upon UV-green irradiation is almost 15%, while for the pattern of period 28 µm there is an intermediate change of ca. 6%.

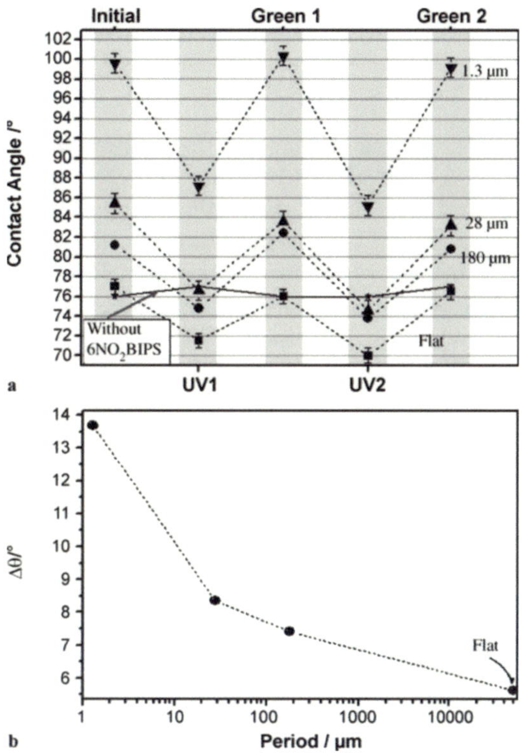

Fig. 9. (a) Mean WCA values of patterned surfaces of 10% wt SP in PEMMA for each pattern, for the flat surface and for the PEMMA surface. The media is taken after studying 10 samples of each case. (b) WCA difference (Δθ) from the initial one after UV irradiation. (λ_{UV}=308 nm, F_{UV}=20 mJ cm^{-2}, λ_{green}=532 nm, F_{green}=25 mJ cm^{-2}) (Lygeraki et al 2008)

Period	1.3µm	28µm	180µm
Initial	0.68	0.88	0.94
UV1	0.80	0.94	0.96
Green1	0.66	0.89	0.91
UV2	0.81	0.94	0.95
Green2	0.68	0.91	0.94

Table 1. Factor *f* calculated by the Cassie–Baxter model under different irradiation conditions.

In conclusion, at this section it is presented the possibility to create both hydrophobic and hydrophilic surfaces starting from the same photochromic polymeric sample by changing the topological parameters of its surface features using soft molding lithography. Due to the photochromic transformations taking place upon alternating UV and green irradiation, these surfaces can reversibly change their wettability. By careful control of the surface topology these changes can be fully controlled and tuned, in such a way that the surfaces can be wetted in a reversible manner.

4. Reversible diffraction efficiency changes of photochromic polymer gratings

4.1 Introduction

The photoinduced reversible volume changes of the lithographically patterned features of such materials are also used for realizing diffraction gratings, operating in the Raman-Nath regime, with controlled and reversible diffraction efficiency (DE). In particular, the alternating pulsed UV and green laser irradiation on the gratings causes the doped photochromic molecules to undergo transformations, which induce reversible dimensional changes to the samples. These volume changes cause reversible changes to the diffraction efficiency of the gratings, which is increased upon UV irradiation and decreased after irradiation with green laser light for various irradiation cycles. The experimental results are confirmed by a theoretical diffraction model. It is proved that the diffraction efficiency changes are attributed exclusively to the reversible dimensional changes of the imprinted structures (Fragouli et al 2008), and not to the refractive index changes as is the case in the majority of previous works. Specifically, most of the gratings with similar thickness as the here presented, are produced after irradiation with interfering beams that cause variation of the refractive index inside the samples mainly consisting of azobenzene polymers and liquid crystals. In these studies the switching procedure in the DE is due to the formation and deletion of the gratings (Yamamoto et al 2001, Fu et al 2005). Moreover, although the thickness of the produced gratings is very small, they exhibit higher DE than most of the volume gratings prepared in this regime (Tong et al 2005, Yamamoto et al 2001, Fu et al. 2005).

4.2 Results and discussion

A grating of period 4 μm and thickness of ca. 240 nm was formed on the photochromic polymeric surface by the SM technique as described at section 2. The experimental setup used for the study of the diffraction efficiency (DE) of the formed grating, is demonstrated in Figure 10. As shown, a continuous wave diode laser operating at λ=822nm, with an incident angle adjusted to have the maximum intensity of the first-order diffracted line (I_1), was used as a reading beam. This wavelength was chosen in order not to be absorbed by the photochromic sample. The zero-order transmitted (I_0) and the I_1 lines, are measured by two photodiodes, and were used for the calculation of the DE (equation 3).

$$DE = I_1/I_0 \tag{3}$$

The DE was calculated for three irradiation cycles, and Figure 11 shows the DE relative change in each case. Initially, the DE of the grating was ca 2.2%, and upon UV laser irradiation it was increasing until it reached a final value. After two minutes in the dark in order to leave time to the system to relax, subsequent irradiation with green laser pulses caused the DE to recover close to its initial value. It is worth noticing that the relative changes of the DE during the first irradiation cycle exhibit big variations between the various examined samples, in contrary with the following irradiation cycles, where the changes are similar for all the examined gratings. This is mainly attributed to internal stresses of the polymer matrix, produced during the preparation of the gratings, that are released in a random way upon irradiation (Liang et al 2007). Nevertheless, at the second cycle during UV irradiation the DE increases with increasing number of pulses, until it stabilizes to an average value of approximately 7.4±3.0% with respect to its initial value. After green irradiation the DE is slowly reaching its initial value with increasing number of pulses. This behavior is repeated also at the third cycle.

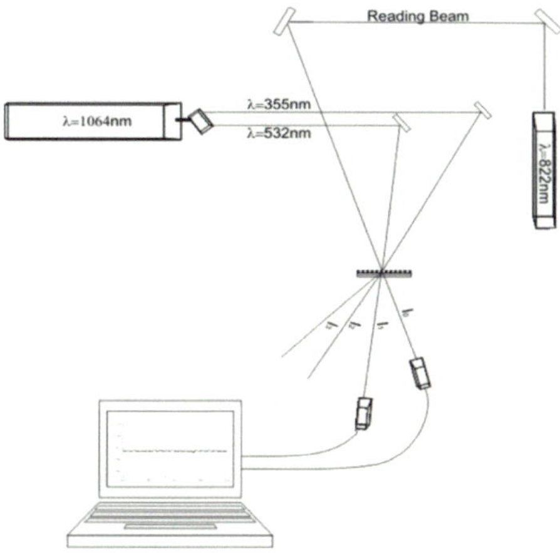

Fig. 10. Experimental setup for the measurement of the diffraction efficiency of the photochromic gratings (10% wt. SP in PEMMA) (λ_{UV}=355nm, F_{UV}=20 mJ cm^{-2}, λ_{green}=532 nm, F_{green}=35 mJ cm^{-2}, $\lambda_{reading\ beam}$=822 nm.)

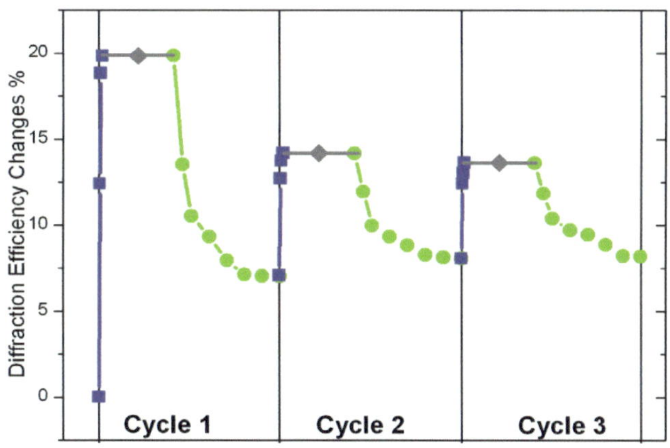

Fig. 11. Diffraction efficiency changes of the grating upon UV-green irradiation. (Fragouli et al 2008)

In order to examine the effect of the refractive index (n) change of the photochromic polymer sample upon UV-green irradiation on the observed change to the DE, ellipsometric measurements on a similar sample were conducted. The results show that before any irradiation for λ=822 nm, the refractive index of the sample is n=1.509. After UV irradiation, the n is higher by Δn=0.029. This difference is very small compared to the periodic refractive index variations in the grating between the photochromic polymer (1.509) and the air (1), which is actually what causes diffraction to occur. Thus it is believed that this change plays a negligible role in the measured DE relative changes. Furthermore, the small thickness of the gratings (ca. 240 nm) reduces the importance of the Δn even more.

Thus, the reversible DE changes can be attributed exclusively to the light-induced macroscopic deformations of the gratings. Specifically, Figure 12 illustrates the reversible macroscopic changes of the grating before and after UV-green irradiation as taken by AFM microscopy. As shown, the width of the stripes of the grating ($\alpha-\beta$) is decreased by ca. 13% after UV irradiation while the distance between the two stripes (β) is increased. A small decrease is also observed in the period of the grating (α) (α and β before UV, 3.971 μm and 2.366 μm respectively; α and β after UV, 3.842 μm and 2.449 μm respectively). After the subsequent irradiation with green light the values recover very close to the initial ones. It is worth noticing that, as shown at Figure 12, there is a dip separating each stripe in two equal parts. As already mentioned at section 2.2, the SM technique which is followed for realizing the grating relies mainly on the capillarity that allows the viscous polymer to spontaneously fill the vertical channels that are made of the recessed features of the elastomeric mold, since the wetting lowers the overall free energy. There is always the possibility that the photochromic polymer may not fill completely such regions, thus may be mostly accumulated in the regions that are adjacent to the protruding areas of the mold, forming thus dips in the central part of the growing capillarity features. This behavior is common to different imprint lithography methods (Zankovych et al 2001, Hong and Lee 2003, Pisignano et al 2004). However, this dip is useful for the AFM morphological analysis of the patterned surfaces, since it makes the volume changes upon UV-green irradiation cycles much clearer. Moreover, it is too narrow to give any contribution to the diffracted light from the grating.

In order to compare the experimental result with the existing theory, the basic equation that describes the intensity distribution of monochromatic light passing through a grating, was used (equation 5). (Born and Wolf 1999)

$$\frac{I}{I_0}=\left[\frac{\sin(\pi\beta p/\lambda)}{\pi\beta p/\lambda}\right]^2\left[\frac{\sin(N\pi\alpha p/\lambda)}{\sin(\pi\alpha p/\lambda)}\right]^2 \tag{5}$$

I and I_0 are the intensities of the light after the grating at various orders of diffraction and at zero order respectively, β is the distance between two successive stripes, α is the period of the grating, N is the number of stripes, λ is the wavelength of the reading beam, and $p = \sin\theta - \sin\theta_0 = m\lambda/\alpha$ (m=0, ±1, ±2 etc) where θ_0 is the angle of incidence and θ the angle of diffraction. The number of the stripes covered by the reading beam was calculated by dividing the diameter of the spot of the beam by the period of the grating in each case. The angle of incidence of the reading beam was θ_0=20°. In each case, by the AFM images it was measured the different value of α and β before and after UV-green irradiation. Taking into account the measured parameters by the experiment and using the equation 5, it was calculated the ratio I_1/I_0, and consequently the DE for the gratings before and after the UV-green irradiation. The theoretical calculations confirm that there is an increase of the DE after the photoisomerization process, by 8.7%. The agreement between the calculated and experimental (ca. 7.4±3.0%) values is notable, taking into account the experimental error due to factors such as the exact value of the laser beam spot size, which leads to an approximate value of the observed stripes, the imperfections of the surface introduced during the grating formation, and so on. The theoretical calculations presented, demonstrate that the decrease of the dimensions of the stripes of the gratings and of the period, are the main parameters that define the change in the DE upon irradiation (Fragouli et al 2008).

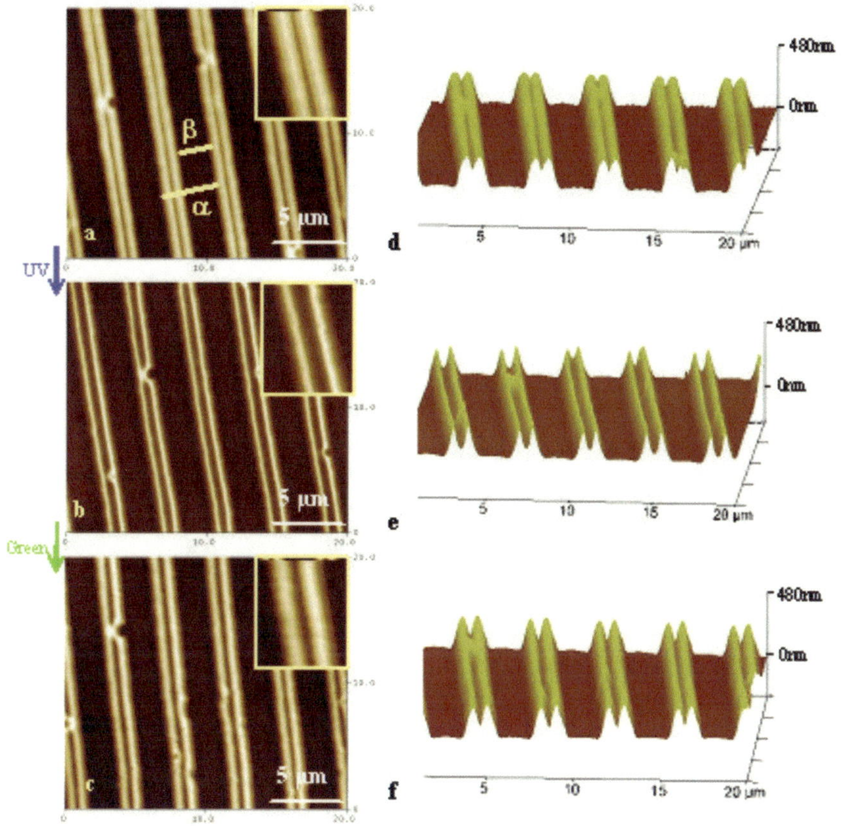

Fig. 12. AFM images of a grating before any irradiation (a) after irradiation with 20 UV pulses (b), and after exposure to 600 green pulses (c). The insets demonstrate a single stripe in magnification. 3D images (d–f) of the grating shown in (a–c), respectively. (α is the period and β is the distance between the stripes, measured at the full width at half-maximum of each feature. (Fragouli et al 2008)

5. Conclusions

In conclusion, it is demonstrated how soft molding lithography can be employed for the preparation of microstructured photochromic polymeric films, which undergo light controlled photomechanical changes responsible for the control of some functional characteristics of the patterned surfaces, namely the wetting properties and the diffraction efficiency. In particular, the light-induced isomerization of the embedded photochromic molecules in the flat surfaces is exclusively responsible for the reversible changes in their wetting properties. When the surface is microstructured by realizing patterns with the SM technique, these wetting properties are greatly enhanced. Moreover the control of the characteristics of the patterns (eg. the period), makes possible to control the light induced alterations in the wetting properties of the structured surface, demonstrating that they are influenced by both the changes in the surface polarity and the volume changes of the patterned structures. Finally, last but not least, it is demonstrated the possibility of fully manipulate the diffraction efficiency of thin photochromic polymer gratings. It is shown, that the produced gratings change their diffraction efficiency in a reversible way upon UV-green laser irradiation. This effect, which is verified also by a theoretical diffraction model, is attributed to the reversible dimensional changes of the imprinted structures, and not to the refractive index changes as is the case in the majority of previous work. Such findings open a way for the production of optically switchable gratings based on reversible dimensional changes. Moreover, the ability to control the wettability of surfaces by microstructuring and to tune it by using photochromic molecules opens the way to the application of these optimized patterns to various microfluidic devices.

6. Acknowledgment

The authors would like to thank Dr. D. Pisignano, and Dr. E. Mele of the Center for Biomolecular Nanotechnologies @UNILE, Istituto Italiano di Tecnologia Via Barsanti, 73010 Arnesano (LE) and Dr. L. Persano of the National Nanotechnology Laboratory (NNL) of CNR - Istituto di Nanoscienze, via per Arnesano, 73100 Lecce, Italy for the formation of the patterned structures presented in this work.

7. References

Athanassiou A.; Kalyva M.; Lakiotaki K.; Georgiou S. & Fotakis C. (2005). All-optical reversible actuation of photochromic-polymer microsystems. *Advanced Materials*, Vol.17, No.8, (April 2005), pp. (988-992)

Athanassiou A.; Lygeraki M. I.; Pisignano D.; Lakiotaki K.; Varda M.; Mele E.; Fotakis, C.; Cingolani R. & Anastasiadis S. H. (2006) Photocontrolled Variations in the Wetting Capability of Photochromic Polymers Enhanced by Surface Nanostructuring. *Langmuir*, Vol.22, pp. (2329-2333)

Athanassiou A.; Varda M.; Mele E.; Lygeraki M. I.; Pisignano D.; Farsari M.; Fotakis, C.; Cingolani R. & Anastasiadis S. H. (2006) Combination of microstructuring and laser-light irradiation for the reversible wettability of photosensitised polymer surfaces. *Applied Physics A*, Vol.83, pp. (351-356)

Athanassiou, A.; Sahinidou, D. ; Arima, V.; Georgiou, S.; Cingolani, R. & Fotakis, C. (2006) Influence of laser wavelength and pulse duration on the degradation of polymeric films embedding photochromic molecules. *Journal of Photochemistry and Photobiology A*, Vol.183, pp.(182-189)

Born M. & Wolf E. (1999) Principles of Optics: Electromagnetic theory of propagation, interference and diffraction of light, 7th Edition, Cambridge University Press.

Brown G. H. (1971) Photochromism, Willey-Interscience, New York
Caprioli L.; Mele E.; Angilè F. E.; Girardo S.; Athanassiou A.; Camposeo A.; Cingolani R. & Pisignano D. (2007) Photochromic spiropyran monolithic polymers: Molecular photo-controllable electroosmotic pumps for micro-fluidic devices. *Applied Physics Letters*, Vol.91, 113113
Cassie, A. B. D. & Baxter, S. (1944) Wettability of porous surfaces. *Transaction of the Faraday Society*, Vol.40, pp.(546-551).
Fragouli, D.; Persano, L.; Paladini, G.; Pisignano, D.; Carzino, R.; Pignatelli, F.; Cingolani, R. & Athanassiou A. (2008) Reversible Diffraction Efficiency of Photochromic Polymer Gratings Related to Photoinduced Dimensional Changes. *Advanced Functional Materials*, Vol.18, pp.(1617–1623).
Fu, S.; Liu, Y.; Dong, L.; Lu, Z.; Hu, W. & Xie, M. (2005). Photo-dynamics of polarization holographic recording in spirooxazine-doped polymer films. *Materials Letters*, Vol.59, pp. (1449-1452)
Fu, S.; Liu, Y.; Lu, Z.; Dong, L.; Lu, Z.; Hu, W. & Xie, M. (2005) Real-time holographic gratings recorded by He–Ne laser in polymer films containing spirooxazine compounds pre-irradiated by UV light. *Optical Materials*, Vol.27, pp.(1567-1570).
Görner H. (1997). Photoprocesses in spiropyrans and their merocyanine isomers: Efects of temperature and viscosity. *Chemical Physics*, Vol.222, pp. (315-329)
Hong, P. S. & Lee, H. H. (2003) Pattern uniformity control in room-temperature imprint lithography. *Applied Physics Letters*, Vol.83, 2441.
K. Ichimura, S.-K. Oh, M. Nakagawa, (2000) Light-Driven Motion of Liquids on a Photoresponsive Surface. Science, Vol.288, pp. (1624-1626).
Kawata, S. & Kawata, Y. (2000). Three -dimensional optical data storage using photochromic materials. *Chemical Reviews*, Vol.100, No.5, pp. (1777-1788)
Liang, G.-D.; Xu, J.-T.; Fan, Z.-Q.; Mai, S.-M. & Ryan, A. J. (2007) Effect of Substrate and Molecular Weight on the Stability of Thin Films of Semicrystalline Block Copolymers. *Langmuir*, Vol.*23*, pp. (3673-3679).
Lygeraki M. I.; Tsiranidou E.; Anastasiadis S. H.; Fotakis, C.; Pisignano D.; Cingolani R. & Athanassiou A. (2008) Controlling the reversible wetting capability of smart photochromic-polymer surfaces by micro patterning. *Applied Physics A*, Vol.91, pp. (397 - 401)
Mitchell, P. (2001) (2001) Microfluidics—downsizing large-scale biology. *Nature Biotechnology*, Vol.19, pp. (717-721).
Patankar, N. A. (2003) On the Modeling of Hydrophobic Contact Angles on Rough Surfaces. *Langmuir*, Vol.19 No.4, pp (1249–1253).
Pisignano D.; Persano L.; Cingolani R.; Gigli G.; Babudri F.; Farinola G. M. & Naso F. (2004) Soft molding lithography of conjugated polymers. *Applied Physics Letters*, Vol.84, No.8, (February 2004), pp. (1365-1367)
Tong, X.; Wang, G.; Yavrian, A.; Galstian, T. & Zhao Y. (2005) Dual-mode switching of diffraction gratings based on azobenzene-polymer-stabilized liquid crystals. *Advanced Materials*, Vol. 17, No.3, pp. (370-374).
Walsha Z.; Scarmagnani S.; Benito-López F.; Abele S.; Nie F.-Q.; Slater C.; Byrne R., Diamond D.; Paull B. & Macka M. (2010) Photochromic spiropyran monolithic

polymers: Molecular photo-controllable electroosmotic pumps for micro-fluidic devices. (2010) *Sensors and Actuators B*, Vol.148, pp. (569–576)

Wenzel, R. N. (1936) Resistance of solid surfaces to wetting by water. *Industrial and Engineering Chemistry*, Vol.28, No.8, pp.(988-994).

Yamamoto, T.; Ohashi, A.; Yoneyama, S.; Hasegawa, M.; Tsutsumi, O.; Kanazawa, A.; Shiono, T. & Ikeda, T. (2001) Phase-Type Gratings Formed by Photochemical Phase Transition of Polymer Azobenzene Liquid Crystal. 2. Rapid Switching of Diffraction Beams in Thin Films. *Journal of Physical Chemistry B*, Vol.105, pp. (2308-2313).

Yu Y.; Nakano M. & Ikeda T. (2003). Directed bending of a polymer film by light. *Nature*, Vol.425, (September 2003), pp. 145

Zankovych, S.; Hoffmann, T.; Seekamp, J.; Bruch, J.-U. & Sotomayor Torres C. M. (2001) *Nanotechnology*, Vol.12, 91. Nanoimprint lithography: challenges and prospects.

Part 5

Extreme UV Lithography

9

Approach to EUV Lithography Simulation

Atsushi Sekiguchi
Litho Tech Japan Corporation
Japan

1. Introduction

1.1 Simulation based on measured development rate measurements

EUV lithography is a reduced projection lithography technology based on 13.5 nm wavelength EUV (Extreme Ultraviolet). Development of EUV lithography is currently underway for the mass production of semiconductor devices for 90 nm design rule applications for ArF dry exposures and for 65 to 45 nm design rule applications for ArF immersion exposures [1-2]. EUV lithography is among the most promising next-generation lithography tools for the 32 nm technology node [3]. The evolving consensus is that EUV exposure technologies will be applied to mass production from the year 2011 [4]. Table 1 showed the relationship among technology node, exposure numerical aperture (NA), and process coefficient factor (k_1) [5]. Achieving the 32 nm node based on an ArF laser source exposure technology will require the development of an optical system with NA increased to 1.55 and k1 improved to 0.26. In contrast, an exposure technology based on an EUV light source will permit the use of an optical system with 0.25 NA for mass production of the 32 nm node with room to spare. The requirement for the k1 factor is an easy-to-meet value of 0.59. These factors underscore the promise and importance of EUV exposure technologies.
However, the development of EUV exposure equipment presents its own set of technology barriers, as does the development of ArF immersion exposure system. A wavelength of 13.5 nm requires a reflecting optical system with a combination of multiple multilayer reflecting mirrors [6], since no lens material can be used in the 13.5 nm wavelength range, if we rule out dioptric lenses. The development of EUV exposure equipment requires further examination of component technologies, including technologies related to light sources, illumination optical systems, projection optical systems, and masks. Although various exposure equipment manufacturers are actively promoting the development of EUV reduced projection exposure equipment [7-8], a resist material for EUV lithography must be developed before the first exposure system can be introduced. We have developed a new virtual lithography evaluation system with lithograph simulation that takes an approach completely different from conventional resist evaluation technologies (direct evaluation method), which require actual patterning to assess resists. The new evaluation system focuses on open-frame exposures using an EUV light source, measurements of development rates at various exposure doses, and lithography simulations based on development rate data. This chapter presents the results of our evaluations of EUV resists using this new system.

Wavelength (nm)	Tech. Node	65 nm	45 nm	32 nm	22 nm	16 nm
	NA	k1	k1	k1	K1	k1
193	0.93	0.31				
	1	0.4				
	1.2		0.28			
	1.35		0.31	0.22	0.15	
	1.35DP			0.20	0.18	
13.5	0.25			0.59	0.41	
	0.35				0.57	0.41
	0.45					0.53

Table 1. Relationship among technology node, numerical aperture (NA), and process factor (k_1)

1.2 System configuration

The virtual lithography evaluation system (VLES) proposed consists of an EUV open-frame exposure system, a resist development analyzer, and a lithography simulator. Fig. 1 is a schematic diagram of the VLES.

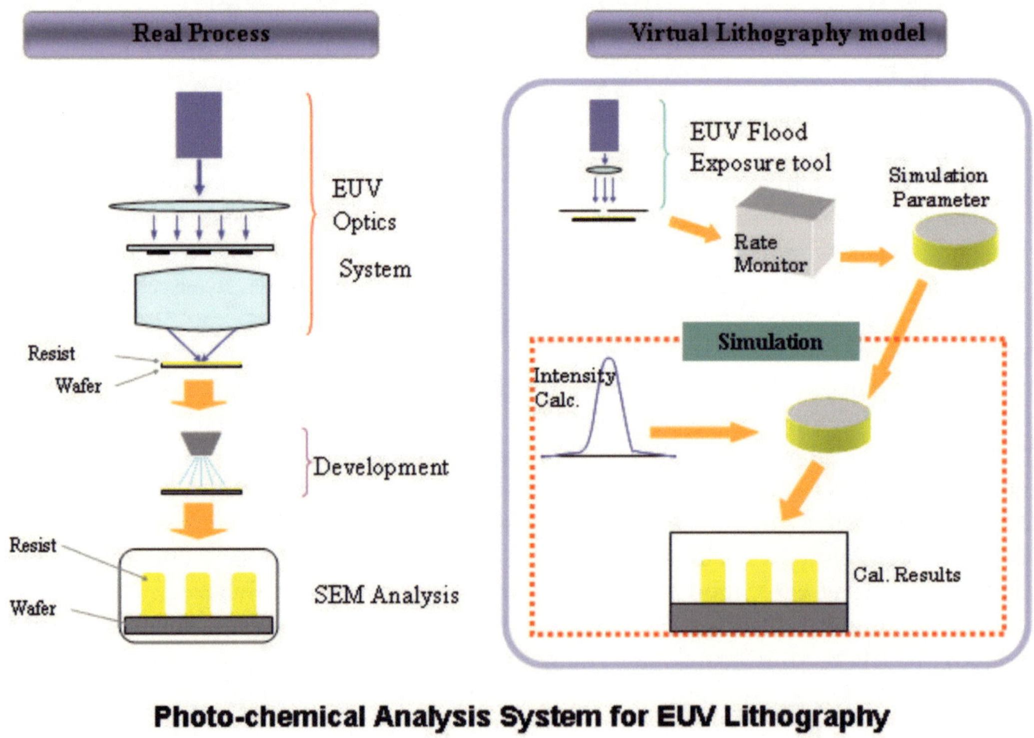

Fig. 1. Schematic diagram of the VLES

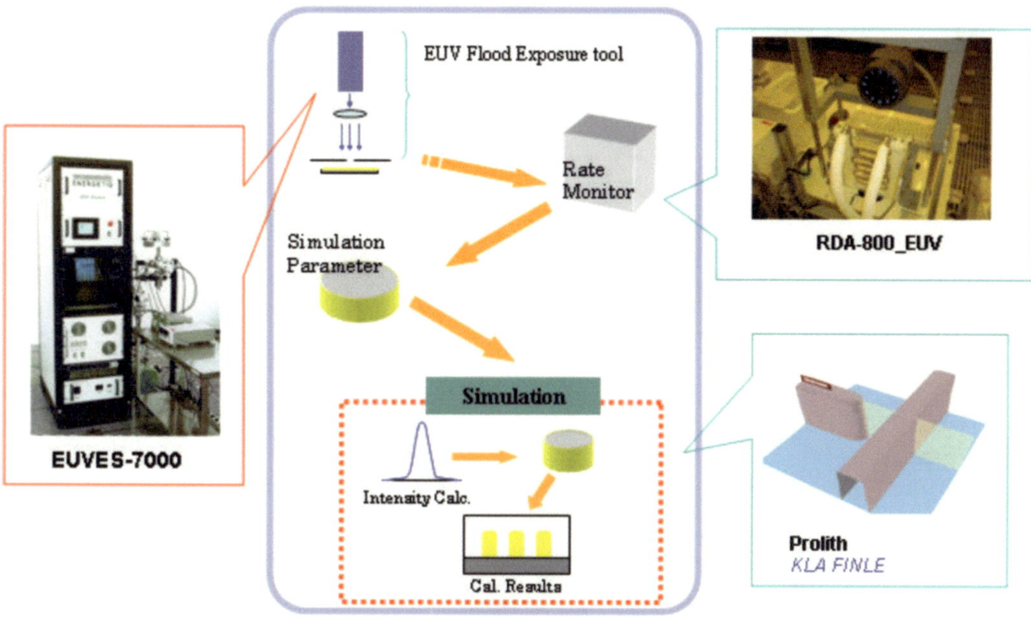

Fig. 2. Analyzers used in the VLES

Fig. 2 shows the analyzers comprising the VLES.

1.2.1 EUV open-frame exposure system (EUVES-7000)

This equipment uses an electrodeless Z-pinch discharge-excitation plasma light source [9] manufactured by Energetiq Technology Inc. It extracts 13.5 nm light using a Zr filter and multilayer reflecting mirrors. The exposure pattern is a 10 mm x 10 mm open frame; 12 exposures can be achieved per wafer at varying exposure doses. Fig. 3 gives an external view of this equipment and a picture of an exposure pattern (after exposure, PEB, and development).

The plasma emissions produced by the EQ-10M pass through the Zr filter to remove UV-region rays. Next, the Mo-Si multilayer reflector selectively reflects only 13.5 nm rays, which are shaped by the aperture into a 10 mm x 10 mm exposure region. The rotary Mo-Si multilayer reflector directs the light at a reflection angle of 45 degrees toward the exposure chamber at the upper section of the equipment during the exposure of a substrate. For power measurements, it rotates and directs the light to the power measurement diode chamber at the lower section of the equipment. Exposures are performed as the wafer rotates. A total of 12 exposures are possible per wafer at varying exposure doses.

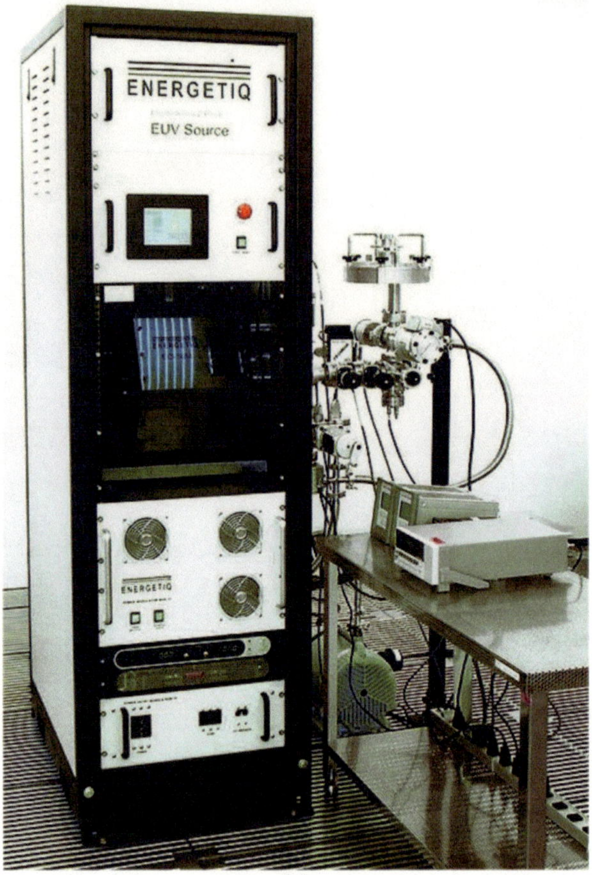

(a)

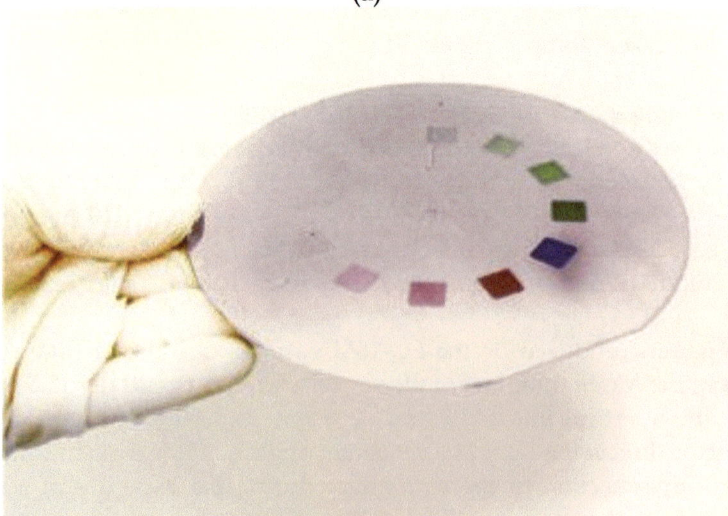

(b)

Fig. 3. (a) External view of EUVES-7000 and (b) exposure pattern

Fig. 4 is a picture of the beam line.

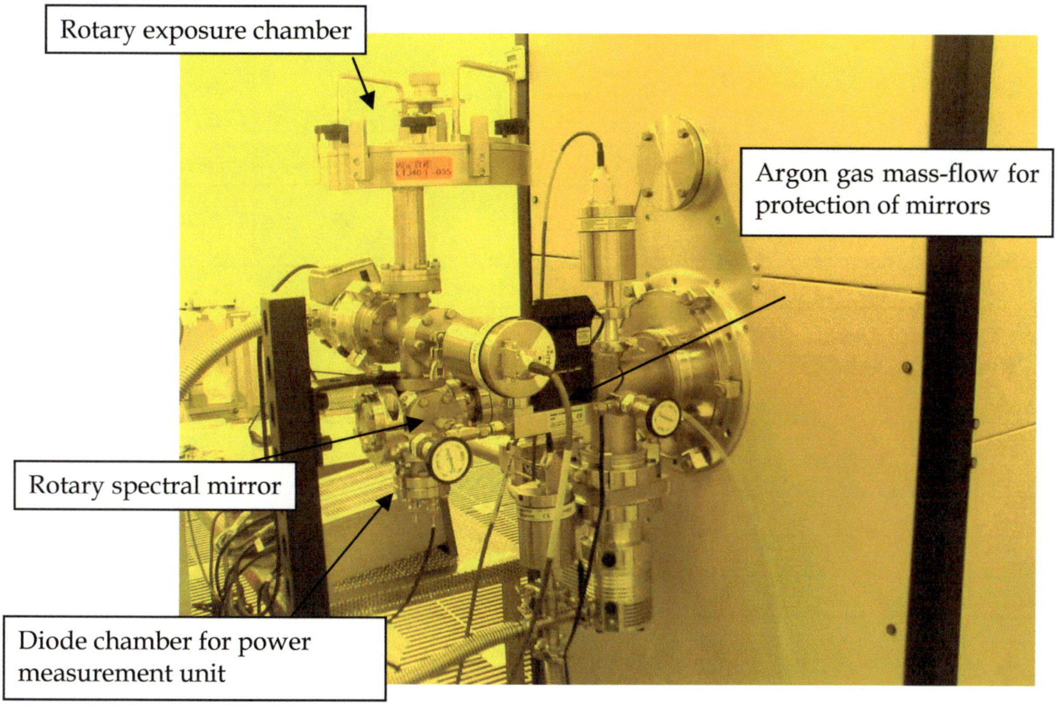

Fig. 4. Beam line for EUV exposure

1.2.2 Resist development analyzer (RDA-800EUV)

Following the exposure, a wafer is processed for PEB. Then, following measurement of film thickness, this resist development analyzer is used to measure the development rate of a resist corresponding to each exposure dose [10].

1.2.3 EUV lithography simulator (Prolith Ver. 9.3)

The obtained development rate data file is imported into the Prolith lithography simulator [11] (manufactured by KLA-Tencor) for EUV lithography simulation.

1.3 Experiment and results

We investigated the sensitivity of positive- and negative-type resists in EUV exposures with the system as described above, then performed simulations using the development rate data obtained.

Table 2 gives the conditions of the resists in our experiment.

The negative-type resists examined were the SAL-601 electron beam resist and SU-8 epoxy-resin-base chemically amplified resist. The positive-type resists used in our experiment were ZEP-520 non-chemically amplified electron beam resist, EUVR-1 and EUVR-2 acrylic-resin-base resists, and EUVR-3 low-molecular-weight resist.

Negative type						
Resist	Maker	Pre-bake		PEB		Thickness (nm)
		Temp. (deg.C)	Time (s)	Temp. (deg.C)	Time (s)	
SAL-601	Rohm & Hass	105	60	115	60	100
SU-8	Nippon Kayaku	90	90	95	100	100

Positive type						
Resist	Maker	Pre-bake		PEB		Thickness (nm)
		Temp. (deg.C)	Time (s)	Temp. (deg.C)	Time (s)	
ZEP-520A	Nippon Zeon	90	90	95	100	100
EUVR-1	TOK	120	90	120	90	100
EUVR-2	TOK	100	90	110	90	100
EUVR-3	TOK	110	90	100	90	100

Table 2. Conditions of resists in the experiment

Fig. 5.(a) shows discrimination curves for negative-type resists; Fig. 5.(b) shows discrimination curves for positive-type resists.

Posi-Type	Eth(60) mJ/cm2	γ60	tanθ
SAl-601	0.928	-1.445	-2.23
SU-8	0.478	-3.023	-3.73

Nega-Type	Eth(60) mJ/cm^2	γ60	tanθ
ZEP-520A	14.710	1.669	1.90
EUVR-1	2.562	2.325	5.06
EUVR-2	8.574	3.997	30.75
EUVR-3	8.497	1.528	14.53

Table 3. Development characteristics

Table 3. shows the results of development characteristic evaluations.
The results show EUVR-2 provides the highest contrast.

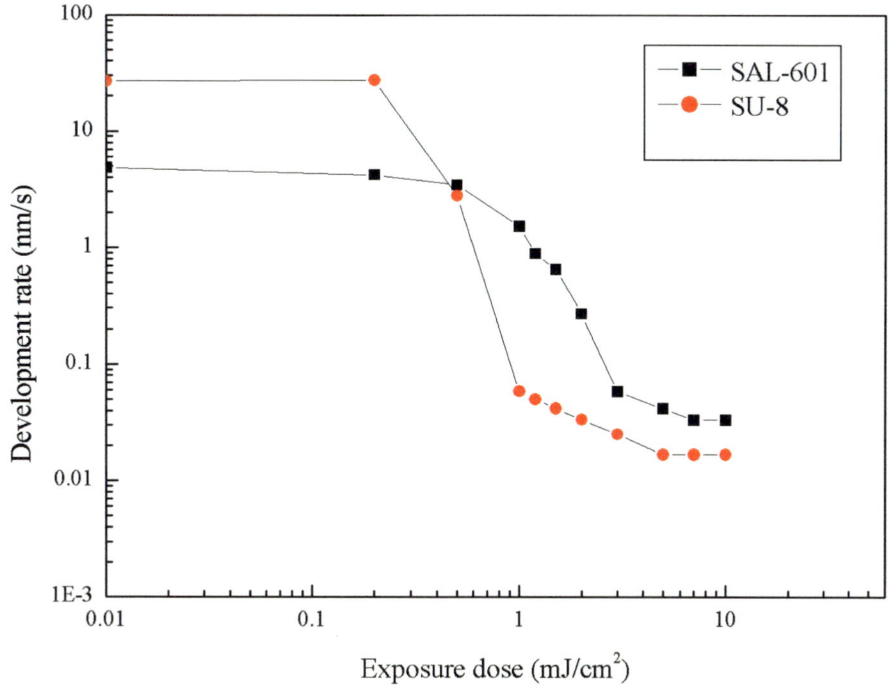

(a) Discrimination curves for negative-type resists

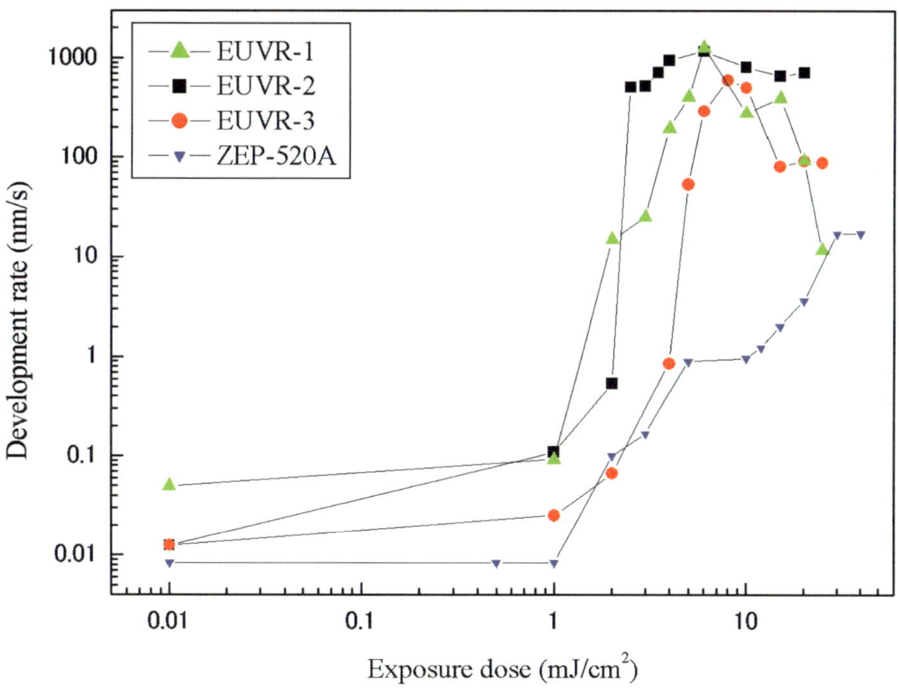

(b) Discrimination curves for positive-type resists

Fig. 5. Relationship between development rate and exposure dose

1.4 Simulation

We performed a simulation using the EUV-PM2 development data. Table 4 gives the simulation conditions.

Wavelength (nm)	13.5
NA	0.3
σ	0.8
Reduction	1/5

Table 4. Simulation conditions (with Nikon HiNA-3)

We examined L&S[*1] patterns and isolated patterns with pattern dimensions of 65, 55, 45, 32, and 22 nm. Defocus was examined using a 32 nm L&S pattern. Figures 6 through 8 show the simulation results. With L&S patterns, resolution can be maintained up to 32 nm. For isolated patterns, the results suggest that resolution on the order of 22 nm is within reach. In the defocus simulation, the simulation results support estimates of an attainable resolution range of -0.1 to +0.1 μm.

[*1] L&S: Line and space

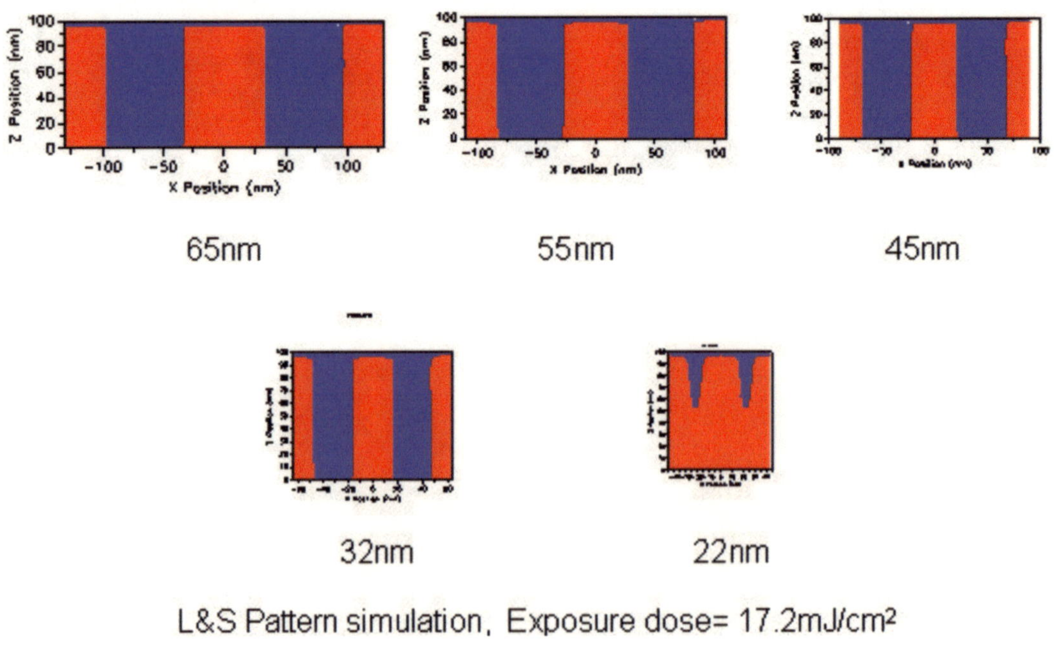

Fig. 6. Simulation results (65-22 nm Line and space patterns)

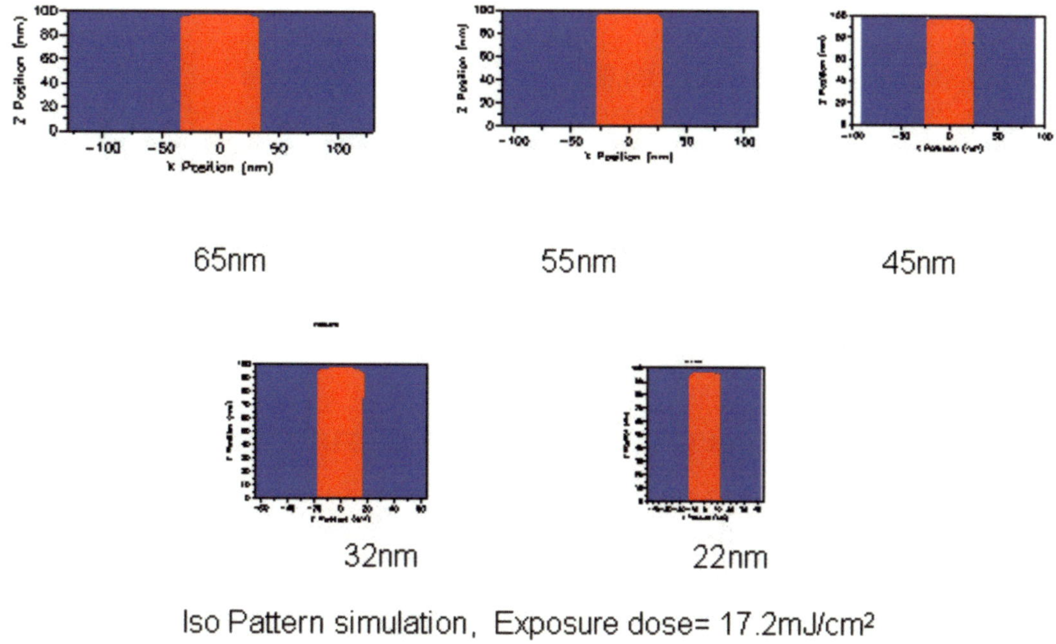

Fig. 7. Simulation results (65-22 nm Isolated patterns)

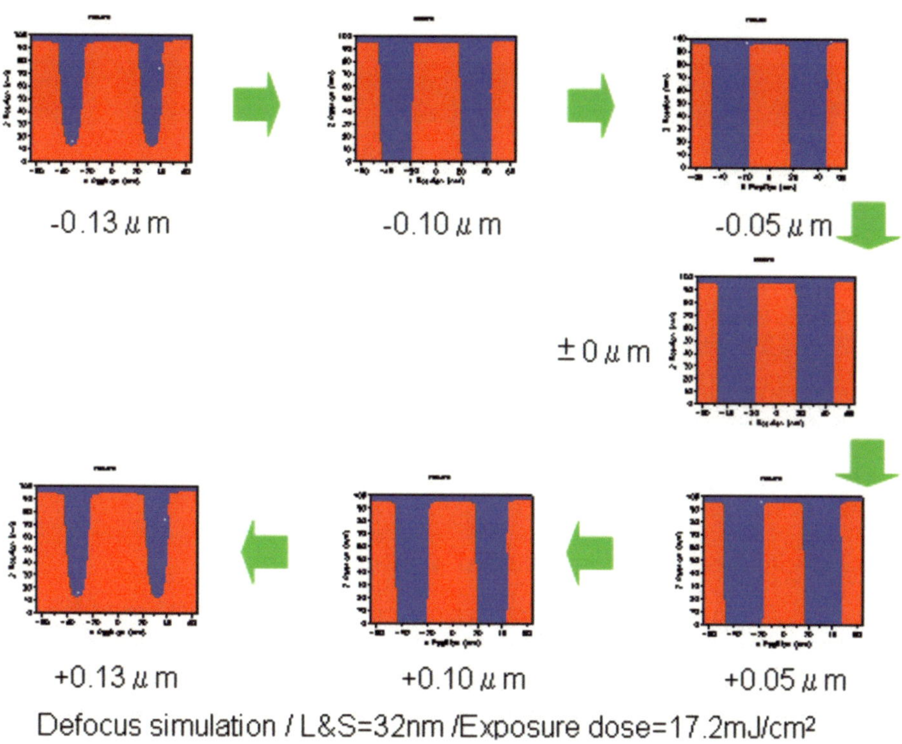

Fig. 8. Simulation results (32 nm L&S/ defocus -0.13～+0.13μm)

1.5 Conclusion

The VLES consists of the EUVES-7000 EUV open-frame exposure system, RDA-800EUV development rate analyzer, and Prolith lithography simulator. We used the VLES to compare the sensitivity and development contrast of negative- and positive-type resists with EUV exposure. We also simulated EUV exposures using development rate data for the EUVR-2, which showed the highest development contrast of all resists tested. The results of the experiment suggest that it should be possible to obtain resolutions of 32 nm with L&S patterns and 22 nm with isolated patterns. We also calculated defocus characteristics with a 32 nm L&S pattern. Based on these calculations, we estimate a focus margin of approximately 0.2 µm in defocus width. We believe using the system as described in this paper will permit the development of photoresist materials for EUV and expedite process development without requiring the purchase of costly EUV exposure equipment.

2. Simulating EUV Resists (Comparison of KrF and EUV Exposures)

2.1 Introduction

According to ITRS Roadmap 2007 Update Version [12], EUVL is currently the most promising candidate for 22 nm half-pitch lithography. The component technologies required for EUVL mass production must be established before the start of mass production of DRAM half-pitch, currently scheduled for 2016. RLS specifications for realizing 22 nm half-pitch resolution were presented at the 7th EUVL Symposium [13] in Lake Tahoe, California, in October 2008.

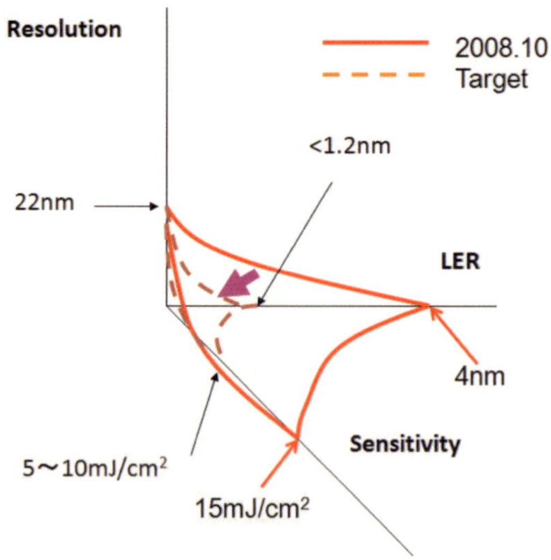

Fig. 9. RLS specifications targeting 22 nm half-pitch

A resolution of 22 nm half-pitch requires sensitivity of 5 to 10 mJ/cm^2 and LER of less than 1.2 nm. At an international conference, it has been pointed out that although resolutions have reached the target value, sensitivity lags, at 15 mJ/cm^2, while LER (Line Edge Roughness) is no less than 4 nm. These are the best values achieved to date. Lithography simulations should prove highly effective in advancing the state of current research, given the time required to perform experiments.

The conventional EUVL simulation method involves obtaining parameters by exposing the resist to EUV. However, EUV exposure equipment is costly, and the types of exposure equipment available are limited. For these reasons, we explored the possibility of performing EUVL simulations using parameters obtained with KrF exposures. The idea was that if we detected no significant differences between parameters obtained with KrF and EUV exposures, we could use the simpler KrF exposure method to obtain valid simulation parameters for EUVL. Using EUV resists, we obtained parameters by performing both KrF and EUV exposures, then compared the parameters and simulation results. This chapter discusses this comparison.

2.2 Simulation parameter measurement system

2.2.1 Exposure equipment for parameter measurement

Fig.10 shows the exposure equipment used in our parameter measurements. The exposure area is an open-frame pattern measuring 10 mm x 10 mm. We used a UVES-2000 for KrF exposures and an EUVES-7000 [14] for EUV exposures.

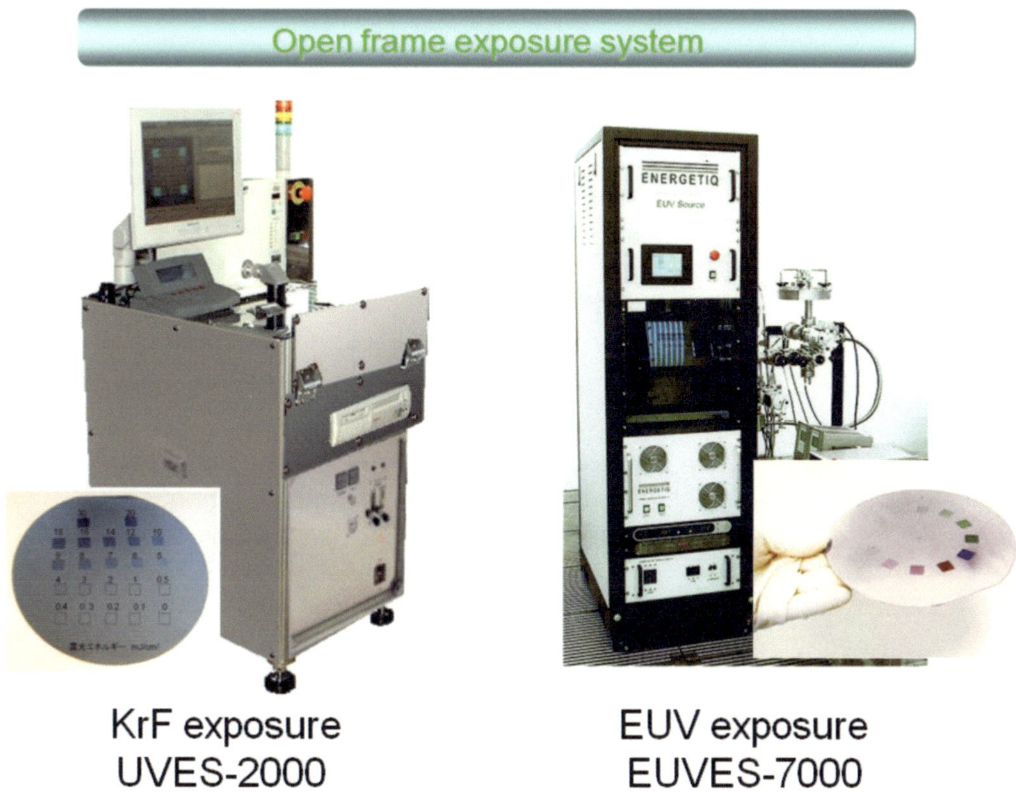

Fig. 10. Open frame exposure tool for KrF and EUV

These exposure tools permit resist exposures on Si wafers and the acquisition of development and PEB parameters.

2.2.2 Development parameter measurement system

We used a development analyzer to measure development parameters. When homogeneous light is irradiated onto a resist film during development, the light waves reflected from the resist surface and light waves reflected from the wafer surface interfere, generating unique waveforms. Analyzing the waveforms of the reflected light allows us to obtain resist development rates. By varying exposure values and measuring resist development rates at different exposures, we can calculate the development parameter, among the simulation parameters [15]. This measurement has been performed before using a monitor wavelength of 470 nm. However, thin films do not generate the interference needed, and a monitor wavelength of 470 nm limits us to resist film thicknesses exceeding 100 nm. Since the film thickness of EUV resists ranges from approximately 50 to 100 nm, we developed a measurement system for our experiments based on a monitor wavelength of 265 nm (Fig.11).

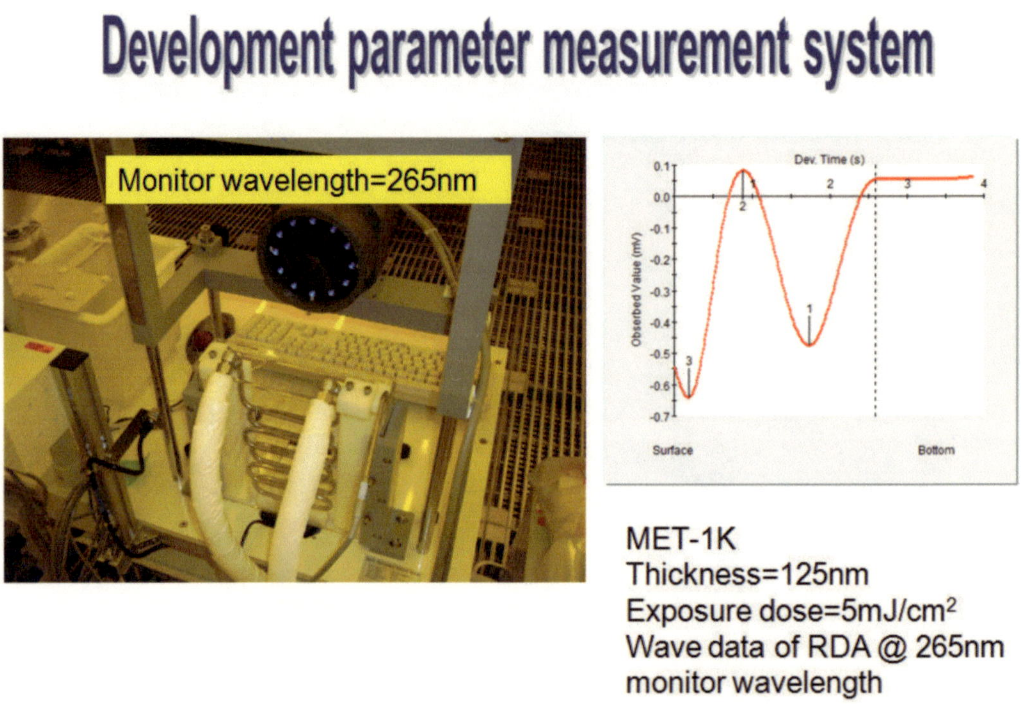

Fig. 11. Resist development analyzer RDA-800EUV

2.2.3 B parameter measurement system

We used the following equation to calculate the B parameter [16-17] of Dill based the resist transmission factor at the time overexposure completely breaks down the PAG.

$$B = -\frac{1}{d}\ln\left(T_{\infty}\right) \tag{1}$$

Here, d is resist film thickness and T_∞ the resist transmission factor at the time overexposure completely breaks down the PAG. We developed a system for measuring the resin transmission factor using EUV light. Incorporating a LPP light manufactured by Toyota Macs as its light source and using a solid Cu target, this system irradiates EUV light onto a Si/Mo multilayer reflecting mirror to measure reflection intensity, while mirror angles are varied. To calculate the spectral transmission factor, we used the difference in reflectance between the case in which resist is applied to the multilayer mirror and the case in which no resist is applied.

Fig.12 illustrates the measurement system and gives a chart of the results of spectral transmission factor calculations for the MET resist.

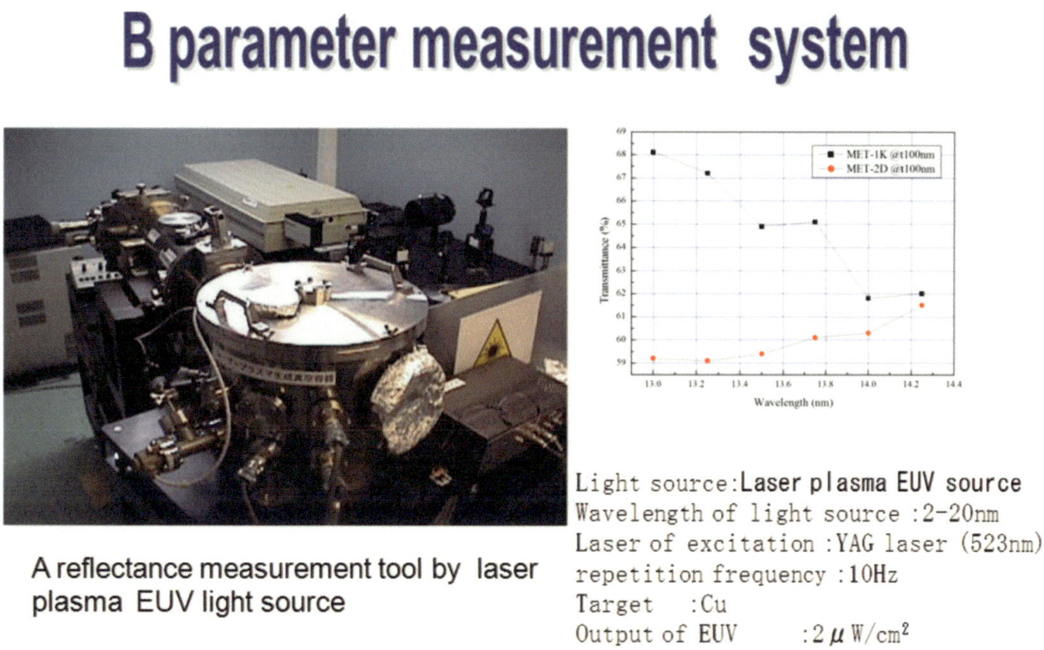

Fig. 12. B parameter measurement system using EUV exposure

2.2.4 De-protection reaction parameter and C parameter measurement system

Fig.13 gives an overview of the PEB parameter measurement system, which exposes resist on an Si wafer using KrF and EUV light. In the next step, we used an FT-IR system with a bake function to plot the de-protection reaction curve while performing PEB. We performed measurements at different PEB temperatures and measured the de-protection reaction parameter by fitting. During the course of fitting, we also obtained the C parameter for Dill. We modified the system [18] to allow irradiation of IR light for measurements on resist film at an angle of 45 degrees and to permit use with a resist film thickness of 50 nm. The resulting system was capable of handling extra-thin resist films ranging from 50 to 100 nm.

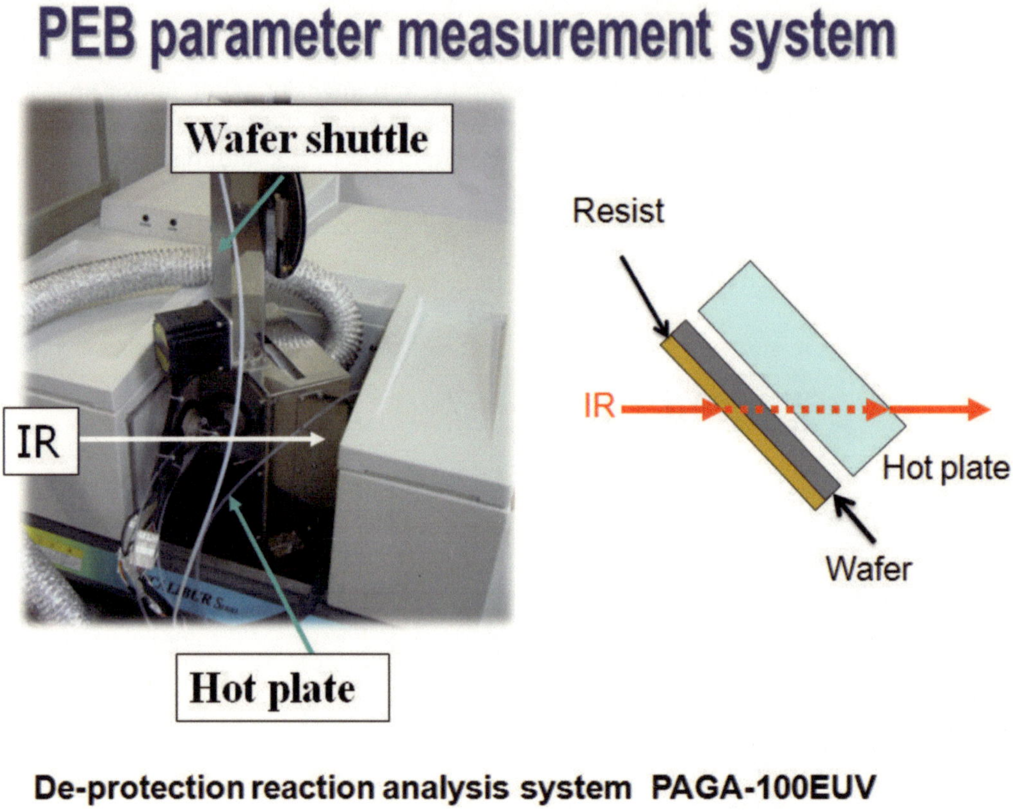

Fig. 13. PEB parameter measurement system

2.3 Parameter measurement results

We measured parameters using EUV chemically amplified resists MET-1K and MET-2D manufactured by Rohm and Haas.
Table 5 gives the process conditions.

	Vendor	Thickness (nm)	PAB		PEB	
			Temp. (deg.C)	time (s)	Temp. (deg.C)	time (s)
MET-1K	R&H	125	130	60	110	90
MET-2D	R&H	125	130	60	110	90

R&H: Rohm and Hass

Table 5. Measurement conditions

2.3.1 B parameter measurement results

Fig.14 shows B parameter measurement results. With KrF exposures, MET-1K and MET-2D yielded values of 0.726 and 0.788, respectively. With EUV exposures, MET-1K and MET-2D yielded 4.32 and 5.21, respectively, indicating greater absorption with EUV exposures.

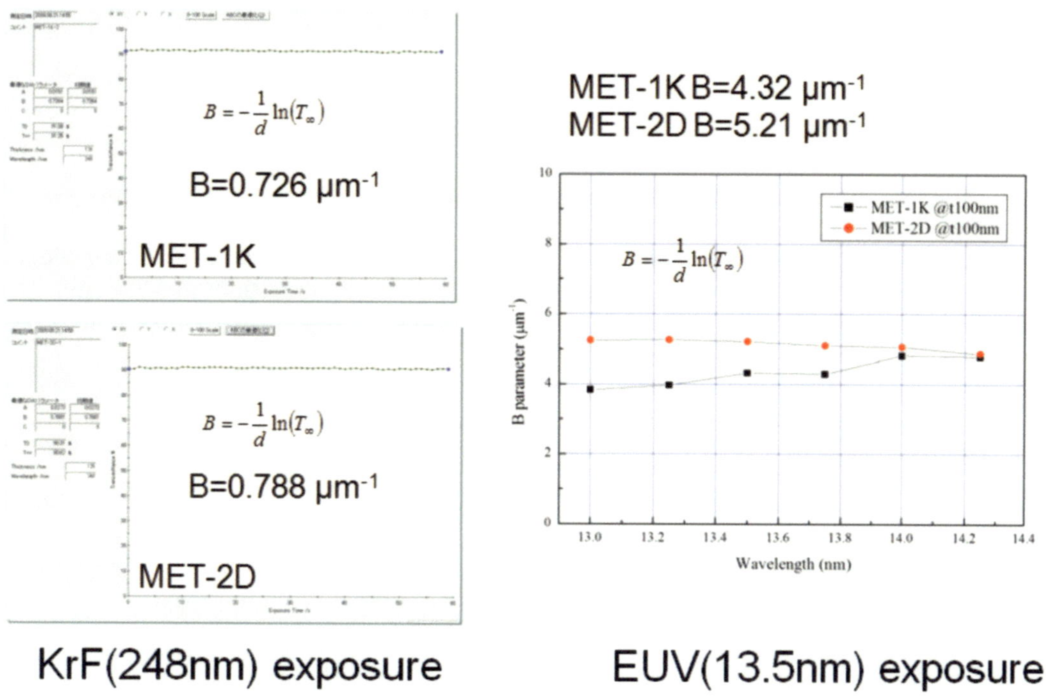

Fig. 14. B parameter measurement results

2.3.2 Development parameter measurement results

Fig.15 compares measurements of the discrimination curve (a logarithmic plot of development rates and exposure values) development parameter. We found no significant differences between development parameter values obtained with KrF and EUV exposures.

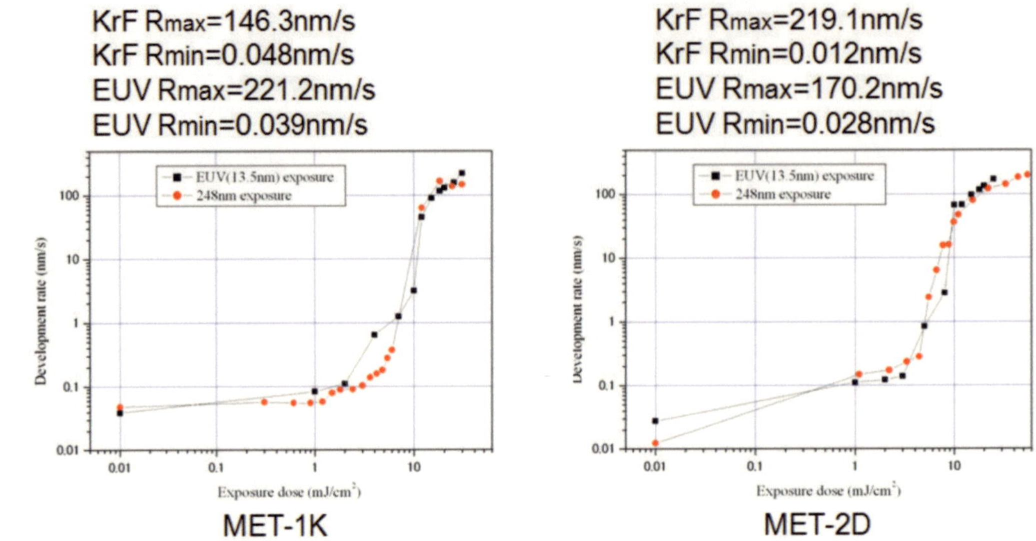

Fig. 15. Comparison of development parameter values

2.3.3 De-protection reaction observations and results of de-protection reaction parameter measurements

Fig.16 shows IR spectra obtained from the MET-1K before and after 16-mJ/cm^2 exposures. The figure indicates weakened (-CO) bonds due to de-protection reactions at 1,230 cm^{-1}. The extent of the peak decline with KrF exposure is roughly identical to that with EUV, indicating the absence of significant differences in de-protection reactions.

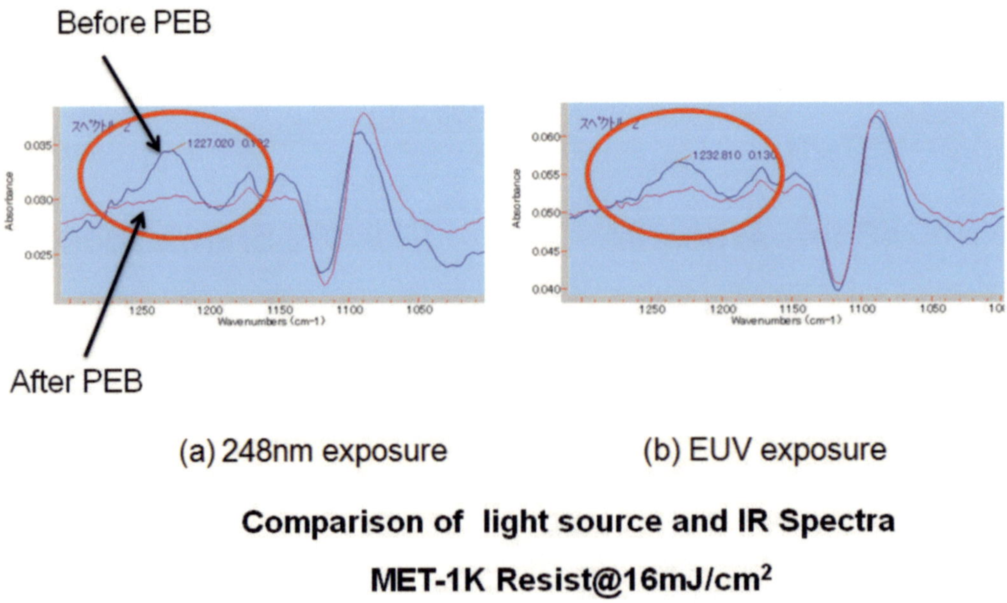

Fig. 16. Observations of de-protection reactions with KrF and EUV exposures

Table 6 is a list of simulation parameter measurement results.

	KrF(248nm) exposure		EUV(13.5nm) exposure	
	MET-1K	MET-2D	MET-1K	MET-2D
ABC-parameter				
A(1/μm)	0	0	0	0
B(1/μm)	0.726	0.788	4.32	5.21
C(cm^2/mJ)	0.017	0.021	0.086	0.090
Development parameters (Mack original model)				
Development R_{max} (nm/s)	146.3	219.1	221.2	170.2
Development R_{min} (nm/s)	0.048	0.012	0.039	0.028
Development M_{th}	0.773	0.783	0.624	0.518
Development n	15.21	17.30	12.65	18.96
PEB parameters				
PEB Amplification Ea (Kcal/mol)	12.21	11.30	3.82	2.23
PEB Amplification LN(Ar) (1/s)	14.89	13.65	5.53	3.55
PEB Diffusion-Controlled Reaction Ea (Kcal/mol)	8.56	4.82	8.80	1.00
PEB Diffusion-Controlled Reaction LN(Ar) (1/s)	10.14	5.54	11.50	1.78
PEB Diffusivity Ea (Kcal/mol)	5.00	5.00	5.00	5.00
PEB Diffusivity LN(Ar) (nm2/s)	23.22	21.24	30.00	30.00

Table 6. Simulation parameter measurement results

2.3.4 Examination of simulation

We performed EUVL simulations using the simulation parameters obtained. Table 7 gives the simulation conditions used.

Simulation condition

NA=0.30

Wavelength=13.5nm

Illumination= Annular (0.3/0.7)

Flare=1%

Line and space=28nm

Resist: MET-1K and MET-2D

Thickness:125nm

Table 7. Simulation conditions

For exposure equipment, our simulation assumed use of the Nikon EUV-1 installed at Selete [19]. Fig.17 shows the simulation results. The indicated exposure value is the exposure level (E_0) that achieved 1:1 resolution from a 28-nm L&S pattern. The development conditions called for 2.38% TMAH and development time of 60 seconds. The quencher diffusion length and PAG diffusion length were set to 20 nm and 10 nm, respectively.

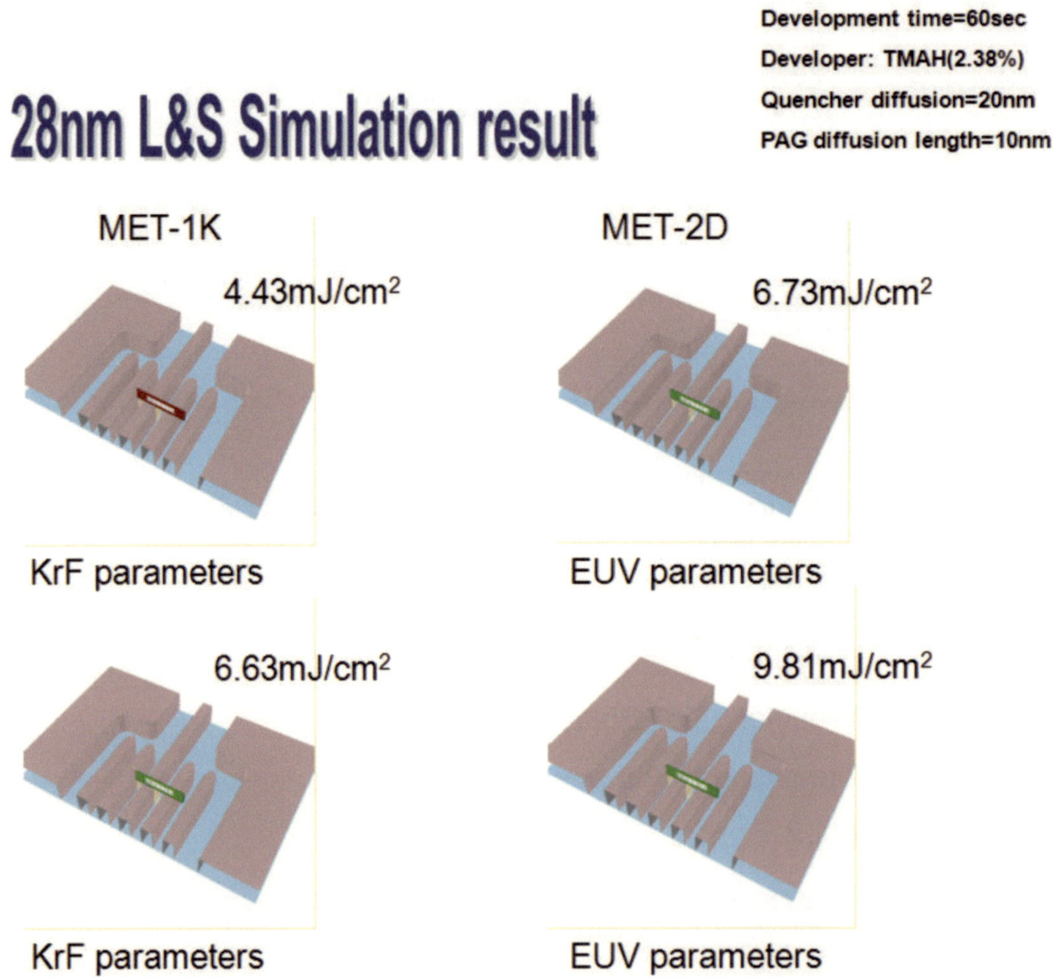

Fig. 17. Simulation results

We compared the results of EUVL simulations based on parameters obtained with KrF exposures to the results of EUVL simulations based on parameters obtained with EUV exposures. While the former simulation results indicated higher sensitivity (approximately 20% higher), we saw no major differences in shape.

2.4 Conclusion

We compared the results of EUVL simulations based on parameters obtained with KrF exposures to the results of EUVL simulations based on parameters obtained with EUV exposures. The former resulted in approximately 20% higher simulation sensitivity, but we saw no major differences in shape. Using parameters obtained with KrF exposure is a roundabout way to perform EUVL simulations. Since EUV exposures in many cases are not readily available, a valid option would appear to be to acquire simulation parameters through KrF exposures and to use these parameters as initial values in calculations for EUVL simulations.

3. References

[1] B. J. Lin, *Proc. SPIE,* 4688, 11, 2002

[2] IMEC 4th Immersion Workshop, September in Belgium, 2005

[3] V. N. Golovkina, P. F. Nealy, F. Cerrina, J. W. Taylor, H. H. Saolak, C. David and J. Gobrecht, *J. Vac. Sci., Technol.* B, 22(1), 99, 2004

[4] Intel H.P.
http://www.intel.com/jp/technology/index.htm?iid=subhdr-JP+tech

[5] ITRS 2005 : International Technology Roadmap for Semiconductors 2005 Edition

[6] H. B. Cao, W. Yueh, J. Roberts, B. Rice, R. Bristol and M. Chandhok. *Proc. SPIE,*5753, 459, 2005

[7] NIKON H.P.
http://www.nikon.co.jp/news/2007/0711_nsr_01.htm

[8] ASLM H.P.
URL : http://www.asml.co.jp

[9] P. Blackborow, *Proc. SPIE,* 6151, 25, 2006

[10] A. Sekiguchi, C. A. Mack, Y. Minami and T. Matsuzawa, *Proc. SPIE,* 2725, 49, 1996

[11] Prolith Version 9.3 User's manual

[12] ITRS LOAD MAP 2007 Up-date in Web

[13] 7th EUVL symposium in Lake Tahoe, CA (2008.10)

[14] A. Sekiguchi, Y. Kono, M. kadoi, Y. Minami,T. Kozawa, S.Tagawa, D. Gustafson and P. Blackborow, *Proc. SPIE,* 6519, 168 (2007).

[15] A. Sekiguchi, Y. Minami, and Y. Sensu, The Electrochemical Society of Japan, Proc. of the 42nd Sysmp. on Semiconductors and Integrated Circuits Technology, Vol. 42, pp. 109-114 (1992).

[16] Dill-B F. H. Dill, W. P. Hornberger, P. S. Hauge, and J. M. Shaw, *IEEE Trans. Electron Dev.,* Vol. ED-22, No. 7, pp. 445-452 (1975).

[17] C. A. Mack, T. Matsuzawa, A. Sekiguchi, and Y. Minami, *Proc. SPIE,* Vol. 2725, pp. 34-48 (1996).

[18] A. Sekiguchi, and Y. Kono, *Proc. SPIE*, 6923, 92 (2008)
[19] H. Oizumi, D. Kawamura, K. Kaneyama, S. Kobayashi and T. Itani, RE-03, 7th EUVL symposium (2008).